Industrial Reliability and Safety Engineering
Applications and Practices

This book addresses the reliability, risk, and safety issues of real industrial systems with application of the latest reliability and risk-based modeling. Related topics such as maintenance decision-making and risk and safety modeling are also addressed with the implementation of decision-making techniques.

The book provides real-life studies on industrial operations along with solutions. It discusses modeling and optimization of reliability and safety aspects in industry and covers reliability maintenance issues in process industries. It goes on to present cost optimization, life cycle costing analysis, and multi-criteria decision making (MCDM) application for risk and safety analysis.

Academic institutions, students, professionals, large companies involved in engineering sciences, research scholars, and investigators working in the domain of Reliability and Safety Engineering and its allied domains will find this book useful.

Advanced Research in Reliability and System Assurance Engineering

Series Editor:
Mangey Ram,
Professor, Graphic Era University, Uttarakhand, India

Modeling and Simulation Based Analysis in Reliability Engineering
Edited by Mangey Ram

Applied Systems Analysis
Science and Art of Solving Real-Life Problems
F. P. Tarasenko

Stochastic Models in Reliability Engineering
Lirong Cui, Ilia Frenkel, and Anatoly Lisnianski

Predictive Analytics
Modeling and Optimization
Vijay Kumar and Mangey Ram

Design of Mechanical Systems Based on Statistics
A Guide to Improving Product Reliability
Seong-woo Woo

Social Networks
Modeling and Analysis
Niyati Aggrawal and Adarsh Anand

Operations Research
Methods, Techniques, and Advancements
Edited by Amit Kumar and Mangey Ram

Statistical Modeling of Reliability Structures and Industrial Processes
Edited by Ioannis S. Triantafyllou and Mangey Ram

Industrial Reliability and Safety Engineering
Applications and Practices
Edited by Dilbagh Panchal, Mangey Ram, Prasenjit Chatterjee, and Anish Kumar Sachdeva

For more information about this series, please visit: https://www.routledge.com/Advanced-Research-in-Reliability-and-System-Assurance-Engineering/book-series/CRCARRSAE

Industrial Reliability and Safety Engineering
Applications and Practices

Edited by
Dilbagh Panchal, Mangey Ram, Prasenjit Chatterjee,
and Anish Kumar Sachdeva

CRC Press
Taylor & Francis Group
Boca Raton London New York

CRC Press is an imprint of the
Taylor & Francis Group, an **informa** business

Designed cover image: Shuttertstock

First edition published 2023
by CRC Press
6000 Broken Sound Parkway NW, Suite 300, Boca Raton, FL 33487-2742

and by CRC Press
4 Park Square, Milton Park, Abingdon, Oxon, OX14 4RN

CRC Press is an imprint of Taylor & Francis Group, LLC

© 2023 selection and editorial matter, Dilbagh Panchal, Mangey Ram, Prasenjit Chatterjee, and Anish Kumar Sachdeva; individual chapters, the contributors

Library of Congress Cataloging-in-Publication Data
Names: Panchal, Dilbagh, editor.
Title: Industrial reliability and safety engineering : applications and
practices / edited by Dilbagh Panchal, Mangey Ram, Prasenjit Chatterjee,
Anish Kumar Sachdeva.
Description: First edition. | Boca Raton, FL : CRC Press, 2023. |
Series: Advanced research in reliability and system assurance engineering,
2767-0724 ; book 13 | Includes bibliographical references and index. |
Summary: "This book addresses the reliability, risk, and safety issues of real industrial systems with application of the latest reliability and risk-based modelling. Related topics such as maintenance decision-making, risk and safety mod0elling are also addressed with the implementation of decision-making techniques. The book provides real-life studies on industrial operations along with solutions. It discusses modelling and optimization of reliability and safety aspects in industry and covers reliability maintenance issues in process industries. The book goes on to present cost optimization, life-cycle costing analysis, and MCDM application for risk and safety analysis. Academic institutions, students, professionals, large companies involved in engineering sciences, research scholars, and investigators working in the domain of Reliability and Safety Engineering and its allied domains will find this book useful"— Provided by publisher.
Identifiers: LCCN 2022038138 (print) | LCCN 2022038139 (ebook) |
ISBN 9780367690311 (hbk) | ISBN 9780367690328 (pbk) | ISBN 9781003140092 (ebk)
Subjects: LCSH: Reliability (Engineering) | Operations research. |
Industrial safety.
Classification: LCC TA169 .I353 2023 (print) | LCC TA169 (ebook) |
DDC 620/.00452—dc23/eng/20221011
LC record available at https://lccn.loc.gov/2022038138
LC ebook record available at https://lccn.loc.gov/2022038139

ISBN: 978-0-367-69031-1 (hbk)
ISBN: 978-0-367-69032-8 (pbk)
ISBN: 978-1-003-14009-2 (ebk)

DOI: 10.1201/9781003140092

Typeset in Times
by codeMantra

This book is dedicated to all editors of this book

Contents

Preface

Reliability Engineering is one of the great concepts which plays an important role in dealing with long-run operation of the real industrial systems. Industrial systems in heavy process industries are highly complex in nature due to their series-parallel arrangement. Due to this complexity, failure prediction is a difficult task for the system analyst who raises serious concerns related to operational safety in the plant. In the past, the world has faced many industrial accidents due to sudden failure of an equipment/subsystem associated with an industrial system. These sudden failures in the plant operation are due to poor maintenance schedule for these systems. To overcome the issue of poor maintenance schedule, it is essential to study and analyze the failure dynamics of the systems. The accuracy of failure dynamics study depends upon the raw data available from the different sources like maintenance experts, computer database, and log book record. Thus, during reliability analysis of a system data, handling is equally important as its mathematical modeling. Further, the concept of risk and safety issues is also very important for enhancing the long-run availability of the system. For risk and safety issues, Failure Mode and Effect Analysis Approach (FMEA) is one of the most popular methods, and various researchers have modified it after a period of time considering the various shortcomings. The application of this approach for studying the risk and safety issues depends upon the reliability parameters of the considered system. Various mathematical theories like fuzzy set theory, type-2 fuzzy set theory, etc. have been incorporated within traditional FMEA approach as per the suitability of the qualitative data obtained from plant's maintenance experts. Application of decision-making approaches within FMEA approach for analyzing risk and safety issues and maintenance decision-making is also very useful in optimizing system's availability over long duration. Considering the importance of all these models in the field of reliability engineering, the aim of this book is to collect cutting-edge research work from the eminent researchers in the domain of reliability engineering, maintenance decision-making, and risk and safety aspects related to different industrial sectors.

Editors

Dr. Dilbagh Panchal is currently working as an Assistant Professor in the Department of Mechanical Engineering, National Institute of Technology Kurukshetra, Haryana, India-136119. He works in the area of Reliability and Maintenance Engineering, fuzzy decision-making, and operation management. He obtained his Bachelor (hons.) in Mechanical Engineering from Kurukshetra University, Kurukshetra, India, in 2007 and Masters (Gold medalist) in Manufacturing Technology in 2011 from Dr B.R. Ambedkar National Institute of Technology Jalandhar, India. He received his PhD in 2016 from the Indian Institute of Technology Roorkee, India. Presently, he oversees three PhD scholars. Seven M. Tech dissertations have been guided by him and two are in progress. He has published 22 research papers in SCI/Scopus indexed journals, and 10 book chapters have been published by him under a reputed publisher. He has edited two books within his field of expertise and has seven books in progress, and he has attended several international conferences. He is a member of various societies like ORSI, Kolkata, ISTE, New Delhi, and IIIE, Mumbai. He is the associate editor of the *International Journal of System Assurance and Engineering Management* (Springer Publication) and regularly reviews many journals such as *International Journal of Industrial and System Engineering* (Inderscience), *International Journal of Operational Research* (Inderscience), *OPSERACH* (Springer Publication), and *Applied Energy* (Elsevier publishers).

Prof. Mangey Ram received a PhD with a major in Mathematics and a minor in Computer Science from G.B. Pant University of Agriculture and Technology, Pantnagar, India. He has been a faculty member for around 14 years and has taught several core courses in pure and applied mathematics at undergraduate, postgraduate, and doctorate levels. He is currently the Research Professor at Graphic Era (Deemed to be University), Dehradun, India, and a Visiting Professor at Peter the Great St. Petersburg Polytechnic University, Saint Petersburg, Russia. Before joining the Graphic Era, he was a Deputy Manager (Probationary Officer) with Syndicate Bank. He is the Editor-in-Chief of *International Journal of Mathematical, Engineering and Management Sciences, Journal of Reliability and Statistical Studies, and Journal of Graphic Era University*; Series Editor of six book series with Elsevier, CRC Press – A Taylor and Francis Group, Walter De Gruyter Publisher Germany, and River Publisher; and the Guest Editor and Associate Editor with various journals. He has over 250 publications with IEEE, Taylor & Francis, Springer Nature, Elsevier, Emerald, World Scientific, and many other national and international journals and conferences. He has also authored/edited over 55 books with international publishers like Elsevier, Springer Nature, CRC Press – A Taylor and Francis Group, Walter De Gruyter Publisher Germany, and River Publisher. His fields of research include reliability theory and applied mathematics. Dr. Ram is a member of many societies, including a Senior Member of the IEEE, Senior Life Member of Operational Research Society of India, Society for Reliability Engineering, Quality and Operations Management in India, and the Indian Society of Industrial and Applied Mathematics. He has been a member of the organizing committee of several international and national conferences, seminars, and workshops. In 2009, he was given the "Young Scientist Award" by the Uttarakhand State Council for Science and Technology, Dehradun, and he has been awarded the "Best Faculty Award" in 2011, "Research Excellence Award" in 2015, and "Outstanding Researcher Award" in 2018 for his significant contribution in academics and research at Graphic Era Deemed to be University, Dehradun, India. Recently, he has received the "Excellence in Research of the Year-2021 Award" by the Honorable Chief Minister of Uttarakhand State, India.

Dr. Prasenjit Chatterjee is currently the Dean (Research and Consultancy) at MCKV Institute of Engineering, West Bengal, India. He has published more than 120 research papers in various international journals and peer-reviewed conferences. He has authored and edited more than 22 books on intelligent decision-making, supply chain management, optimization techniques, and risk and sustainability modeling. He has received numerous awards including Best Track Paper Award, Outstanding Reviewer Award, Best Paper Award, Outstanding Researcher Award, and University Gold Medal. Dr. Chatterjee is the Editor-in-Chief of the *Journal of Decision Analytics and Intelligent Computing.* He has also been the Guest Editor of several special issues in different SCIE/Scopus/ESCI (Clarivate Analytics) indexed journals. He is also the Lead Series Editor of Smart and Intelligent Computing in Engineering (Chapman and Hall/CRC Press); Founder and Lead Series Editor of Concise Introductions to AI and Data Science (Scrivener-Wiley); AAP Research Notes on Optimization and Decision Making Theories and Frontiers of Mechanical and Industrial Engineering (Apple Academic Press, co-published with CRC Press, Taylor and Francis Group); and River Publishers Series in Industrial Manufacturing and Systems Engineering. Dr. Chatterjee is one of the developers of two multiple-criteria decision-making methods called Measurement of Alternatives and Ranking according to COmpromise Solution (MARCOS) and Ranking of Alternatives through Functional mapping of criterion sub-intervals into a Single Interval (RAFSI).

Dr. Anish Sachdeva is working as a Professor in the Department of Industrial and Production Engineering at Dr B.R. Ambedkar National Institute of Technology Jalandhar, Punjab, India. He obtained his B.Tech. (Industrial Engineering) from Regional Engineering College, Jalandhar (now known as National Institute of Technology, Jalandhar) in 1994; M.Tech. (Industrial Engineering) from Punjab Technical University, Jalandhar in 2003; and PhD (Department of Mechanical and Industrial Engineering) from IIT Roorkee in 2008. He has published more than 100 research articles in international journals and conferences of high repute. He has guided 60 M.Tech. and 12 PhD candidates in pursuing their PhD degree. His academic life includes serving as a peer reviewer in journals, acting as session chair in many international conferences, and conducting several training programs. His areas of interest include Reliability and Maintenance Engineering, Advanced Machining, Supply Chain Management, Stochastic Modeling, and System Simulation. He has also organized five international conferences at NIT Jalandhar as Organizing Secretary and Convener.

Contributors

Ankur Bahl
School of Mechanical Engineering
Lovely Professional University
Phagwara, India

Nand Gopal
Department of Industrial and production
Engineering
Dr B.R. Ambedkar National Institute of
Technology
Jalandhar, India

Hemalata Jena
School of Mechanical Engineering
KIIT Deemed to be University
Bhubaneswar, India

Mukesh Kumar Jha
Department of Information Technology
Dr B.R. Ambedkar National Institute of
Technology
Jalandhar, India

Mohit Kumar
Department of Information Technology
Dr B.R. Ambedkar National Institute of
Technology
Jalandhar, India

Pawan Kumar
Department of Mathematics
Central University of Haryana
Mahendragarh, India

Rakesh Kumar
Department of Mathematics
Shaheed Bhagat Singh State University
Ferozepur, India

Sachin Kumar
Department of Mathematics
KIET
Ghaziabad, India

Sudhir Kumar
Department of Production and
Industrial Engineering
National Institute of Technology
Kurukshetra, India

Dinesh Kumar Kushwaha
Dr B.R. Ambedkar National Institute of
Technology
Jalandhar, India

Munish Mehta
School of Mechanical Engineering
Lovely Professional University
Phagwara, India

Bijaya Bijeta Nayak
School of Mechanical Engineering
KIIT Deemed to be University
Bhubaneswar, India

Rajlaxmi Nayak
Mechanical Engineering Department
JK Lakshmipat University
Rajasthan, India

Dilbagh Panchal
Department of Mechanical Engineering
National Institute of Technology
Kurukshetra, India

Richa Pandey
Graphic Era Hill University
Haldwani, India

Parul Punia
Department of Mathematics
Central University of Haryana
Mahendragarh, India

Amit Raj
Department of Mathematics
Central University of Haryana
Mahendragarh, India

Navneet Rana
Department of Mathematics
Guru Nanak College
Punjab, India

Rajaram Rout
Global Institute of Management
Bhubaneswar, India

Anish Sachdeva
Department of Industrial and
    Production Engineering
Dr B.R. Ambedkar National
    Institute of Technology
Jalandhar, India

Jitendra Kumar Samriya
Department of Information Technology
Dr B.R. Ambedkar National Institute of
    Technology
Jalandhar, India

Sanjay Sharma
Department of Applied Sciences and
    Humanities
Ajay Kumar Garg Engineering College
Ghaziabad, India

P.C. Tewari
Department of Production and
    Industrial Engineering
National Institute of Technology
Kurukshetra, India

Anand Tyagi
Department of Applied Sciences and
    Humanities
Ajay Kumar Garg Engineering College
Ghaziabad, India

1 ISO Tank Containers for Inland Transportation of Petroleum

Safety Review in Indian Perspective

Hemalata Jena
KIIT Deemed to be University

Rajaram Rout
Global Institute of Management

Rajlaxmi Nayak
JK Lakshmipat University

Bijaya Bijeta Nayak
KIIT Deemed to be University

CONTENTS

DOI: 10.1201/9781003140092-1

1.1   INTRODUCTION

Various liquids and gases are required to be produced, stored, and transported to meet essential needs of people to make their life more comfortable and luxurious. Storing the highly corrosive, volatile, flammable fluid is challenging. Hence, a guideline must be followed for safely storing and handling these highly hazardous materials [1,2]. Its storage container is designed in such a way to avoid any fatal accident during handling and transportation. Containerized transportation of these materials helps in avoiding several loading and unloading steps while changing the modes of transportation between rail, road and sea encountered between the place of origin and the destination location. The transportation route of these containers involves different environment conditions, a variety of service fluids, and may pass through various countries having different legal frameworks. The containers need to comply with all requirements of such frameworks. The prevailing capabilities of the ports, railways or other agencies involved in the multimodal transportation need to be assessed, and their prior consents are essential for handling/mounting/demounting/transportation of such hazardous cargo in the International Organization for Standardization (ISO) containers. Hence, the container manufacturers have to adopt a standard design specification for the intermodal transport of various goods. They follow specifications and codes given by ISO to maintain high standard of quality control and safety provisions of the container for longer service life and to meet stringent statutory safety regulation. Hence, these is called ISO containers. These containers can hold a huge array of different goods and can be used in all modes of transportation, such as trucks, train or ships. Hence, integrity of their design and structure plays an important role in their transportation. For example, they are designed in such a way to overcome in extremely adverse saline environments and to survive in collisions of ships during water transport [3]. They should possess sufficient structural integrity for lifting by cranes or other heavy machinery during movement and for carrying flammable and dangerous fluids and gases at high pressure like petroleum and compressed natural gas. Petroleum products are widely used in our daily life ranging from cooking (liquefied petroleum gas) to fuels (petrol, diesel, gasoline, compressed natural gas, etc.) in the transportation industry. It is categorized under dangerous goods, but their use makes our life easier and comfortable. Handling, storing and transportation of these items in ISO tanks is a science that helps the manufacturer to avoid their damage. Accidents occur owing to non-adherence to stipulated safe operating procedures during loading/unloading operations, human error, etc. These accidents have the potential to cause loss of several human lives and demolish hard-earned resources causing a major worry for the disaster management authorities. Chang et al. [4] have studied 242 accidents related to petroleum storage tanks. It is found that 33% of accidents are due to human errors. There are many studies conducted to evaluate the property losses in these industries and to find the root causes of these accidents [5–7]. It is found that 85% accidents are due to fire and explosion.

Safe transportation and storage of crude oil and petroleum products are vital in an oil and gas industry. In its unrefined state, crude oil is transported by two primary modes: tankers and pipelines. Risk estimation of transportation of petroleum has received much attention in different modes of transportation such as pipelines

[8], railway [9] and road [10,11]. Transportation of petroleum products by road, truck or rail has a high risk of accidents. Hence, precaution has to be taken during storing, handling and transportation. From different studies and reports, it is observed that the shipments of oil by rail bring about a great risk of accidents [12–15]. In India and worldwide, the best safe mode for transportation of petroleum is pipeline [16,17].

The movement of crude oil starts from the oil fields to petroleum refineries to storage areas, where the crude oil is refined to get different products. Petroleum products are transported from the refinery to market and distribution locations in the country through truck tanker, rail wagons or pipelines. The transportation and storage of petroleum is done by oil-marketing companies and other private operators. Hence, the transportation and storage industry becomes a very complex system due to the association of many independent private or government owners [18]. Such complex scenario makes rigorous inspections, stringent standards of compliance, and regulations generated from industry initiatives and from government mandatory. The safety, efficiency, tanker hull strength and pipeline integrity have become the most important parameters to avoid fires, oil spills and oil leaks. For this, Petroleum Act 1934 [19] and Petroleum Rules 2002 [20] are introduced to specify various provisions with respect to safety concerning the import, transport, storage, refining and blending of petroleum. The provisions of Petroleum Rules 2002 are formed in such a way that the human intervention in the handling of petroleum is reduced to avoid the episode of accidents involving petroleum in the world's second-most populous country India, where one accident can cause loss of several lives due to high population density. These statutory provisions specify the in-built safety designs. Schedule III under Petroleum Rules 2002 and IS:13187:1991 [21] exclusively specify the design requirements, the test procedures and maintenance philosophy of petroleum road tankers for transport of petroleum in bulk on land. Presently, the road transport of the petroleum products is only by the road tankers with the tank fixed on the truck chassis. The ISO tank container can be an alternative to the petroleum road tankers which does not require to be permanently fixed on the truck chassis and can be removed from truck to rail or ship at ease. In Indian scenario, ISO tank containers are not used for road transport of petroleum owing to inadequate handling infrastructure and safety concerns. These are also not permitted under Petroleum Rules 2002 for inland transportation of petroleum. The design difference between the petroleum road tankers and ISO tank containers is detailed in this chapter from a safety point of view.

1.1.1 ISO CONTAINER TYPES

The tank container is used for the safe transportation and storage of fuel and gas. Different modes of transportation of ISO tank containers are through land (truck or railway), water (ship) or air (airplane). All pressure vessels containers (tank) should be certified by the ISO. Tank containers are generally referred to as "portable tanks" or "ISO tanks".

There are several varieties of ISO containers depending on their mode of operation and use [8].

- **Tank containers:** It contains cylindrical pressure vessels mounted on the frame, mostly double-walled, resting on a rectangular rigid steel frame (ISO frame).
- **Flat racks and flat platforms:** To transport the heavy equipment.
- **Insulated or thermal shield containers:** To transport frozen or cold goods.
- **Open-top type containers:** The top side used to be open in these containers, and they are suitable to fill the coal, minerals or grains from the top side.
- **Cube containers:** To fill the general-purpose goods from one side (open side). It is enclosed on all other sides.
- **Reefer or refrigeration containers:** To transport perishables items like food materials. It is installed with an on-board refrigeration unit.

The present study on ISO tank container is conducted for enlightening its safety precaution. These containers are given a unique number prescribed by Bureau International Containers (BICs) as per ISO 6346 [22]. To support and protect the ISO tank, it is covered with a steel frame (ISO frame) which is also ISO certified. The frame design follows strict standards including longitudinal and lateral inertia. To make possible lifting and stacking, corner castings are provided in the ISO frame. To reduce the material handling and transportation cost, all the containers must have mechanically secure frames of the same size, similar filling/discharge connections and locking arrangements. Tank design is an important factor for its performance, safety during handling, storage and transportation. The tank design follows international regulations and regional regulatory standards including UN, IMDG, ADR, RID and CFR49 [23]. These regulatory bodies permit design approval and prototype testing, inspection and testing of the ISO tank independently.

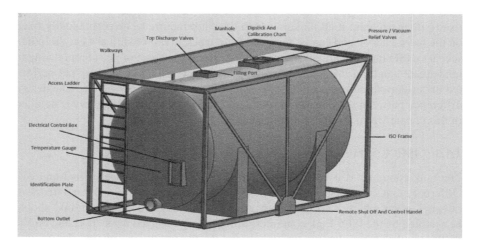

FIGURE 1.1   Cylindrical pressure vessels mounted on a rectangular rigid steel frame.

ISO tank container consists of two main parts: one is the outer structural frame (ISO frame), and the other is the inner tank made of steel to hold the liquid or gas as shown in Figure 1.1. Various configurations of valves and fittings are added during the manufacturing of ISO tank. ISO tank has a maximum gross weight of 36 metric ton. The various accessories/safety fittings of ISO tank containers are discussed below [24,25]:

- **Identification plate:** A plate having common information on technicalities and origin of the ISO tank is displayed in the tank. This is referred to as identification plate or certificate plate.
- **Label holder:** The label holder is available in the tank for displaying the safety labels required to recognize the tank contents.
- **Manhole:** It is a hole for entering into the tank for internal inspection and maintenance. It is a hole of about 500 mm in diameter.
- **Top discharge valve:** Provision for pumping out the material from the top of the tank.
- **Filling port:** It is a nozzle for loading of the material into the tank.
- **Bottom outlet:** Outlet nozzle fitted with a valve for pumping out the material from the bottom of the tank.
- **Electrical control box:** To control all electrical supply to electrical heating coils, remote shutoff control, etc.
- **Remote shutoff control handle:** For emergency closure of the bottom outlet foot valve.
- **Walkways:** To access various valves, it is placed on top of the tank container.
- **Ladder:** It is placed at the rear of the container.
- **Dip stick:** It is generally used to measure liquid level in the container.
- **Document tube:** The document tube is watertight and secured area to avoid document stealing during transport.
- **Safety fittings:** Pressure relief valve is termed as safety fittings to discharge excess pressure in case of pressurized containers.
- **Temperature gauge:** To regularly check the temperature of liquid/gas being transported.
- **Emergency shutoff valve:** To close the discharge of tank content during an emergency.
- **Pressure gauge:** Pressure-measuring device to monitor the pressure of fluid inside the tank.
- **Heating coils:** To maintain the desired temperature of the congealing liquids during transportation. The high pour point petroleum products like furnace oil and bunker oil require heating coils.

1.1.2 MATERIALS FOR ISO TANK CONTAINERS

Steel, stainless steel, aluminium and fibre-reinforced polymer (FRP) are used to fabricate the tank container. Different materials are used for different corrosive liquids and gases. The details are discussed below:

- **Steel:** Steel is used for corrosive chemicals such as Class 8 dangerous goods. But due to low corrosion resistance, it limits its applications. But stainless steel has high corrosion resistance which gives it more suitability. Marine-grade stainless steel used for inner shells of the tank container has high corrosion resistance. This steel shell protects the cargo. Cautious design of stainless steel tank containers is essential to make the container leak-proof, which appreciably reduces the risk of spillage during transportation and loading/discharging processes. Present-day ISO frame is commonly available in the standard size of 20′ long, 8′ wide and 8′6″ high [25]. The containers are designed to suitably fit into this frame. ISO frame of different sizes are also available to fit the specific requirement. A liner is used to protect the steel tank from highly corrosive chemicals. This liner is generally made from thermoplastic polymer like rubber and polyethylene.
- **Aluminium:** Aluminium is used for non-corrosive gas or liquid such as Class 3 (flammable) dangerous goods.
- **Fibre-reinforced polymer:** FRP is also used in making tank container. Due to its high specific strength and high corrosion resistance, it excites the researchers to work on different fibre polymer composites whose properties can be manipulated as per the requirement. FRP composite material also is lightweight and has high chemical resistance to carry the fluid in the tank without failure [26]. Among all fibres, carbon fibre is generally used for making ISO tank containers. It exhibits excellent corrosion resistance compared to conventional steel tanks. Carbon FRPs are used for making Omni Tank manufactured as per ISO specifications [27].

1.1.3 CONTAINER SAFETY CERTIFICATE (CSC) OF ISO CONTAINERS

The frame of ISO containers is designed conforming to ISO-1496-3:1995 code, and the safety provisions for their coastal movement are tested by agencies approved by the IMDG (International Maritime Dangerous Goods) [28]. As per the guidelines of the international convention for safe containers, all the ISO containers should have a container safety certificate (CSC) plate affixed on it which is issued by the manufacturer, and this certificate is renewed every 30 months by an authorized inspector after checking its integrity. The CSC number is the unique identification number given to each tank container.

1.1.4 DESIGN SPECIFICATION OF ISO CONTAINERS

ISO tank container has a particular design specification which helps the tank manufacturer to adapt to the stringent safety standards of that particular code and conforming to all the test procedures of the pressure vessel design code. For example, volume of the container is 10, 20, 30 and 40 ft, water capacity is ranging from 10 to 51 kL and design pressure is ranging from 7 to 34.4 bar as per design codes conforming to different national and international standards (ASME Sec. VIII Div I & Div 2, IS: 2825, EN-13530 or AD2000) are commonly available in the international market for transportation [29].

1.2 SAFETY FITTINGS INSTALLED IN THE PETROLEUM ROAD TANKERS AVAILABLE IN INDIA

Safety fittings such as emergency shutoff valve, emergency vent, pressure vacuum valve, and vapour recovery equipment are fitted in the oil tankers to avoid the accident. There are also other safety precautions available as per Schedule III of Petroleum Rules 2002 [20] and IS:13187:1991 [30]. These are as follows:

i. **Restriction on the permissible capacity of tank:** A tank having a net capacity beyond 5 kL is divided into compartments by oil-tight partitions. Each compartment should not have net petroleum capacity beyond 5 kL. If the aggregate carrying capacity of the petroleum tanker or a tank semi-trailer exceeds above 25 kL, then the maximum permissible capacity of each is increased to 7 kL. However, the maximum net carrying capacity of the petroleum tank truck or tank semi-trailer is restricted to 40 kL.

ii. **Multi-compartment philosophy:** Multi-compartment model reduces the axle load and helps in lowering the centre of gravity to keep the trailer in stable condition. When a large-capacity trailer carrying petroleum, the instability of the trailer happens due to fluctuations of liquid in all direction causes major change in the centre of mass. This results in major changes in the axle load, which affect the stability of the truck. This multi-compartment model also helps in avoiding catastrophic explosion if the tank is encountered with fire incident.

iii. **Pressure vacuum (PV) valve and manhole:** Each compartment is fitted with a manhole to facilitate human entry into the tank for inspection, repairs and maintenance. The petroleum road tankers are designed to operate at atmospheric pressure. However, there will always be the generation of additional vapours owing to the increase in atmospheric temperature. In order to release these additional vapours generated during the transportation and discharge of excessive vapours in exigencies, each compartment manhole cover is fitted with various safety fittings. The pressure vacuum (PV) valve is one of the main mechanisms of the safety fittings which is placed in the manhole cover. The purpose of the valve is to regulate the pressure inside the tank automatically. The PV valve also guards the tank and diminishes the volatilization loss of volatile oil.

iv. **Vapour recovery system:** The prevailing environment regulations mandate the prevention of volatile organic compounds from escaping into the atmosphere. The tank is offered with a vapour recovery system whose function is to condense the petroleum vapour for their reuse. The vapour recovery system has various valves like oil vapour recovery valve, bottom valve and fuel outlet valve. Vapour recovery valves and vents are also designed to direct the passage of vapours to the correct location for their recovery so that it cannot create a pressure build-up in the road tankers during loading and unloading.

v. **Emergency venting system:** Each compartment of the petroleum road tanker is fitted with an emergency vent. It is provided with either pressure-actuated

vent or fusible vent or a combination of both at an urgent situation to vent the
fire during fire exposure. However, in case of tanks of capacity 25 kL and
above, tank is not to be offered with fusible vent.

vi. **Shear section:** Each compartment is offered with the bottom valve provided
with a shear section. The shear section is meant to break when exposed
to severe strain in case of external impact on the discharge faucet. Upon
impact, the shear section breaks and the outlet valve is detached from the
tanker. It is also designed to close the discharge from the tanker upon break-
ing of the shear section.

vii. **Emergency cut-off valve:** The outlet of each compartment is fitted with
emergency cut-off valve attached with the fusible link to stop discharge of
petroleum products in case of vehicle exposed to external fire and the ambient temperature reaching above 93°C. The fusible alloy is melted at this
temperature, and the emergency cut-off valve ceases the outflow of tank
content.

1.3   COMPARISON OF THE DESIGN REQUIREMENTS OF PETROLEUM ROAD TANKERS AND ISO TANK CONTAINERS

Road tankers must meet government regulation. The manufacturer puts forward
slightly different equipment to meet different specifications, but it must meet Schedule
III of Petroleum Rules 2002 [20]. On the other hand, ISO tank containers are designed
and fabricated as per the provisions of ASME Sec. VIII, Div I & Div 2; IS: 2825;
EN-13530; AD2000, and mounted on standard ISO frame of 10 ft/20 ft/40 ft size.

The comparison of the design requirements of petroleum road tankers as specified in Schedule III of Petroleum Rules 2002 with that of typical ISO/portable tank
containers is discussed in the following:

I. **Materials of construction of the tank shell:** The third schedule of Petroleum
Rules 2002 specifies the mild steel to be used for the construction of the tank
shell. The ISO container can be constructed using a variety of materials like
FRP/mild steel/stainless steel, which are suitable to specific service fluid
properties.

II. **Tank loading type:** The petroleum road tanker used in India is of top-load-
ing/bottom-loading type depending on the tank configuration. Each method
has its uniqueness.

Top loading: Filling fuels or liquid from the top side of the container. It is
unsafe and risky for fluids like gasoline because it evaporates into the
atmosphere easily when loaded from the top which results in the explosion at the filling station. Besides, the evaporated gas reduces the total
volume of the cargo and increases the time of filling.

Bottom loading: As compared to top loading, it is safe, efficient and less
time-consuming. It is recommended for all types of fuel and liquid
cargo. As per the environment policy, all the petroleum tankers in the
country are being converted to bottom-loading facility with vapour

recovery system, which may not be feasible for ISO containers. Even if the vapour connection nozzle is provided on the ISO containers, they are usually of flanged type, they will not fit into the present available loading infrastructure in the petroleum storage depots and their introduction would require major modification and dedicated bays to be provided in the loading gantries.

III. **Requirement of division of tanks into compartments:** For the petroleum tankers in the country, a tank having a net capacity exceeding 5 kL should be divided into more than one compartment within the tank by oil-tight partitions. Each compartment should have a net capacity below 5 kL and the maximum permissible carrying capacity of tank container is limited to 40 kL as per the Petroleum Rules 2002. In case tank truck or tank semi-trailer is having a net carrying capacity above 25 kL, then the maximum permissible capacity of each compartment is 7 kL. This division of the tank into separate airtight compartments prevents sympathetic ignition of the petroleum in adjacent compartments if one compartment catches fire. On the other hand, the ISO tank containers are mostly single compartment for the entire tank. Even partition plate is available in some customized designs; they are like baffle plate, and airtight segregation is not available. Hence, if one ISO container filled with petroleum catches fire, it is more likely that the entire tank content will burn. This is a significant drawback of ISO containers with respect to safety in comparison to the petroleum road tankers used in India.

When the tank truck is running on a slope or applies break, the liquid in the tank fluctuates in all directions. In large-capacity fuel tankers or trailers, the fluctuations cause major changes in the centre of mass. Such problem is more prevalent when the tank or its compartments are half filled or partially filled. Such instability can cause accidents resulting in the toppling of the tank trucks or ISO tank-mounted trailers. The petroleum road tanks having airtight compartments already reduce the fluctuating liquid force, thereby maintaining the vehicle stability. But the ISO containers need to have baffles to withstand such problems.

IV. **Redundancy in safety fittings:** The petroleum road tanker is provided with a set of separate safety fittings and inlet/outlet nozzles for each compartment of the tank. But the ISO tank containers are not provided with airtight compartments for positive isolation; instead, such containers are provided with baffles which are partly open having common vapour space and a single manhole and single bottom valve for the entire tank. In case of failure in malfunction or failure of any of the safety fittings or valves, the entire tank content will be exposed to disasters. Therefore, such tanks/containers are required to be manufactured and maintained with the highest level of safety standards. The likelihood of any disaster is the combination of the frequency of occurrence and the quantum of the flammable material involved. The provision of airtight partition in petroleum road tankers reduces the quantum of the material involved in the disasters when compared with ISO containers.

V. **Tare weight of the tank:** The 20 ft ISO tank containers usually have a water capacity of 24–26 kL with a shell thickness of 6 mm [31]. The shell thickness of similar capacity petroleum road tanker is 3.5 mm. Hence, the unladen weight of the ISO tank container of 26 kL capacity and its ISO frame is much higher than the unladen weight of 26 kL petroleum road tanker with its accessories. This will result in significantly higher fuel consumption for road transport of petroleum in ISO tank containers when compared with the similar capacity petroleum road tankers. Hence, the transportation cost in ISO container is higher if multimodal transportation is not involved along the route from source to destination.

The other significant design differences between the petroleum road tankers and the ISO containers are listed in Table 1.1

TABLE 1.1

Comparison of Design Parameters of ISO Container and the Third Schedule of Petroleum Rules, IS:13187:1991

S. No.	Design Parameters	Tank of Road Tanker	Typical ISO Tank Containers
1.	Thickness of shell material	2 mm thickness for tank to 3.3 mm	The shell thickness is calculated as per the design code depending on the design pressure and tank diameter
2.	Operating pressure	Atmospheric pressure	1.74 bar for typical T3 ISO tank container
3.	Testing of tank integrity	Hydrotest of the tank at 0.316 kg/cm²g	Hydrotest as per design code which is minimum 1.3 times the design pressure
4.	Emergency control system	Remote-operated foot valve with fusible link	Remote-operated foot valve with fusible link
5.	Normal venting facility provided on the tank	Pressure vacuum valve which will open at 0.21 kg/cm²g pressure limit and on reaching a vacuum of 5 cm water gauge inside the tank	Pressure relief valve
6.	Emergency venting in case of shell being exposed to external fire	Fusible-type emergency vent which will open on exposure to temperature not exceeding 93°C. The emergency vent net area in square centimetres = 8 + 4.3 * (the gross capacity of the compartment in cubic metres)	Rupture disk with port size fixed as per the design code
7.	Specification for design of tank vehicle	Driver cabin at minimum 15 cm gap from the tank head	Any trailer with load-carrying capability more than the ISO tank container weight can be used

(Continued)

TABLE 1.1 (*Continued*)
Comparison of Design Parameters of ISO Container and the Third Schedule of Petroleum Rules, IS:13187:1991

S. No.	Design Parameters	Tank of Road Tanker	Typical ISO Tank Containers
8.	Weld joint efficiency	Joint efficiency >85%	As per design code which is usually more stringent
9.	Tank gauging arrangements	Dip pipe	Dip pipe or level gauge
10.	Top filling pipe extension to bottom of tank to avoid static electricity	Top filling pipe up to bottom	Top filling pipe up to bottom
11.	Tank anchoring with the chassis	Securely anchored with the chassis by U-bolts	To have corner casting, secured by twister locks for fixing on any trailer chassis and easier removal
12.	Over turn protection of tank and safety fittings	Provision available for overturn protection	The entire tank is housed inside the ISO frame. The ISO frame can withstand up to nine high stacking in fully loaded condition and hence adequately protected for overturning
13.	Marking and identification of the tank	Making as per norms, there is no unique no. for the tank. Vehicle no. is the identity of the tank	ISO tanks are assigned global unique no.
14.	Maximum tank capacity	Maximum 40 kL	ISO tanks of water capacity up to 44 kL are available mounted on 40 ft ISO frame

Source: Data taken from Refs. [20,21,24,30–32].

1.4 REDUCTION IN LOADING/UNLOADING BY ISO CONTAINERS

The majority of fire incidents involving the petroleum road tankers happen during loading the tanks or during unloading the tank content from the tankers. If the number of times of loading and unloading is reduced, the likelihood of fire incidents and resulting disasters can be minimized. For example, if some speciality petroleum products are to be transported from a refinery in Haryana to user location in an island or a foreign country, this would require loading of the product into a road tanker at the refinery and transporting it to the nearest port. The tank content will be unloaded at the petroleum storage installation at the port. Then, the same product will again be pumped in pipelines to the dedicated oil jetty in the port for loading into a ship. The ship will travel to the nearest port of the user location and unload the petroleum products in the port. One more road tanker will take the petroleum product to the user location. Such type of movement involves several loading and unloading operations, thereby increasing the likelihood of accidents. If the quantity of the service liquid is less, transportation in pipeline and ship is not economically viable. The product-handling losses are more, and the chances of product contamination are high. The ISO containers are best suited for such cases. Once the petroleum products are loaded into the ISO tank, it can be transported by road or rail to the nearest port, and the ISO tank can be loaded into any cargo ship at any jetty in the port. The ship carries the ISO tank to the nearest port of the user, and from the port, the tank is transported by rail or road to the user premise for unloading. Hence, there is only one time loading into the ISO tank and one time unloading from the ISO tank. The resulting reduction in the number of loading/unloading into the tank can significantly reduce the likelihood of accidents.

1.5 ADDITIONAL SAFETY PRECAUTIONS TO BE OBSERVED DURING LOADING AND UNLOADING AND LIFTING OPERATIONS OF ISO CONTAINERS

Additional safety precautions need to be undertaken during loading and unloading and lifting operations of ISO containers. Independent third-party inspection should be conducted to ensure safe mounting of ISO container on the vehicle chassis for inland movement. The whole operation of lifting and mounting of ISO tank containers on the chassis of vehicles and also the demounting from the chassis at the recipient locations should be supervised by a competent technical supervisor. The existing petroleum storage installations have gantries designed to load/unload petrolium products into road tankers and may not be suitable for ISO tank loading/unloading. These petroleum depots or terminals require a provision for enough open area in their gantries owing to the large turning radius of trailers on which ISO containers are mounted. The vehicles used for transportation of ISO container should not be allowed to change during transit to the user destination. Such ISO containers should not be set as static storage tanks as this would compromise the safety requirements of the storage facility.

The following safe handling precautions should be observed at the time of lifting and handling of ISO tank containers:

- The Material Safety Data Sheet (MSDS) of the fluid being transported is read cautiously to identify the hazards related to handling of the fluid.
- The shell material in contact with the service fluid should be compatible with its physical and chemical properties. ISO containers should be selected accordingly.
- Suitable crane having adequate load-bearing capacity should be used to lift the containers and the lifting force should be applied vertically to the four top corner fittings of the ISO frame so as to avoid any additional external loading on the tank.

- A suitable method should be adopted for loading and unloading the petroleum products into the ISO tank by considering its risks like explosion hazard, flammability, volatility, accumulation of static electricity etc.
- The road worthiness of the prime mover should be ensured before mounting the petroleum-filled ISO container.
- The ISO tank containers are not allowed to be dragged or slid over any surface so as to prevent damage to the vessel.
- All the equipment engaged in the handling of such ISO containers filled with inflammable material should not be allowed beyond its permissible safe working capacity.

1.6   TRAINING OF TANKER CREW

Tanker designer has to take safety precaution through the adoption of several provisions given by government regulation. That precaution has several factors like the design of the tankers, which is discussed earlier, safety provisions attached to road tankers, and, most importantly, training to drivers' for carrying highly flammable materials and plying through densely populated areas [33]. Collisions of tank containers can lead to death or sever disability of people; in addition, these can cause environmental pollution and financial burdens to both society and the individuals involved [34]. One significant cause that's often overlooked is the competency and skill level of drivers and crew members. The main reasons of accidents by the driver are consumption of alcohol, eating while driving, high speed and rash driving, inadequate sleep of the drivers, violation of traffic rules, carelessness, fatigue and drowsiness due to long journey, driving at night, bad weather conditions and making phone calls while driving. In addition to the driving skills, the crew is also trained in handling unforeseen emergencies during their journey to avoid an accident. Emergencies that can occur while handling are significantly different between ISO containers and road tankers owing to their different design and operating parameters. The petroleum road tanks operate at atmospheric pressure, whereas the ISO containers are designed to withstand pressurized storage. Introduction of the ISO containers will require substantial enhancement of skills of the crew members.

1.7   SANITATION REQUIREMENT OF ISO TANK
CONTAINERS BEFORE REUSE

If the ISO tank containers are used for transportation of different service fluids in different trips, they must go through sanitation procedures before reused [35]. The sanitization process may vary from washing to chemical cleaning depending on the requirement of the tank content. They should be cleaned and again certified by ISO for its reuse. It may cost a few hundred dollars to the recipient of a shipment.

1.8   SCOPE FOR ISO TANK IN INDIA

The manufacturing of ISO tank containers is yet to pick up in India, even though huge numbers of these ISO containers are imported to the country every year for transport of speciality gases and chemicals [36]. Due to geographical compulsions, these ISO tank containers are being used for the transport of Aviation Turbine Fuel (ATF), Liquefied Natural Gas (LNG), and other speciality volatile chemicals to islands of India like Andaman and Nicobar, for which large-scale infrastructure is not economically viable due to low demand. ISO tank containers being portable and rugged in the design are best suitable for these conditions.

1.9 CONCLUSIONS

In view of the foregoing discussions regarding the variation in the design philosophies of the petroleum road tankers and that of the ISO tank container, it is inferred that the petroleum road tankers are safer for inland transportation due to their airtight partitions and compartment wise separate set of safety fittings. Hence, these road tankers should continue to be used as primary means for bulk transportation of the petroleum by road in prevailing Indian conditions.

The ISO tank containers are more suitable for transportation of speciality petroleum products that are traded in limited volume and involve multimodal transportation along the route, and adequate storage installations are not available. However, these ISO tanks should have baffles when the tank capacity exceeds 7.0 kL, and provisions for vapour recovery should be present to avoid emission of volatile organic compounds to the environment. The interchanging of trailer chassis should be avoided during road transport due to inadequate infrastructure availability. The ports do not need dedicated oil jetties, and ISO containers filled with petroleum products can be loaded or unloaded to ships in any container terminal. Hence, the introduction of ISO tank containers will greatly help the petrochemical and other related industries.

REFERENCES

1. Basheer A, Tauseef SM, Abbasi T, Abbasi SA. A template for quantitative risk assessment of facilities storing hazardous chemicals. *International Journal of System Assurance Engineering and Management.* 2019;10(5):1158–72.
2. Wang D, Liao B, Zheng J, Huang G, Hua Z, Gu C, Xu P. Development of regulations, codes and standards on composite tanks for on-board gaseous hydrogen storage. *International Journal of Hydrogen Energy.* 2019;44(40):22643–53.
3. Tan X, Tao J, Konovessis D. Preliminary design of a tanker ship in the context of collision-induced environmental-risk-based ship design. *Ocean Engineering.* 2019;1(181):185–97.
4. Chang JI, Lin CC. A study of storage tank accidents. *Journal of Loss Prevention in the Process Industries.* 2006;19(1):51–9.
5. Marsh & McLennan. *Large Property Losses in the Hydrocarbon Chemical Industries: A Thirty-Year Review* (17 ed.). New York: M & M Protection Consultants, 1990.
6. Marsh & McLennan. *Large Property Losses in the Hydrocarbon Chemical Industries: A Thirty-Year Review* (17 ed.). New York: M & M Protection Consultants, 1997.
7. Marsh & McLennan. *The 100 Largest Losses 1972–2001: Large Properties in the Hydrocarbon-Chemical Industries.* New York: M & M Protection Consultants, 2002.
8. Citro L, Gagliardi RV. Risk assessment of hydrocarbon releases by pipelines. *Chemical Engineering Transactions.* 2012;20(28):85–90.

9. Saat MR, Werth CJ, Schaeffer D, Yoon H, Barkan CP. Environmental risk analysis of hazardous material rail transportation. *Journal of Hazardous Materials*. 2014;264:560–9.
10. Tomasoni AM, Garbolino E, Rovatti M, Sacile R. Risk evaluation of real-time accident scenarios in the transport of hazardous material on road. *Management of Environmental Quality. 2010; 21(5): 695–711.*
11. Fabiano B, Currò F, Reverberi AP, Pastorino R. Dangerous good transportation by road: From risk analysis to emergency planning. *Journal of Loss Prevention in the Process Industries.* 2005;18(4–6):403–13.
12. Brown M. (2017). *APNews Break: Thousands of Defects Found on Oil Train Routes, US News and World Report* (accessed 20 June 2020). https://apnews.com/17ff7868111b4ecd9455e42e48ba101a.
13. Lowy J. *Here's Why So Many Oil Trains Have Derailed This Year. Business Insider.* Business Insider, Inc. 2015:10.
14. Green KP, Jackson T. (2017) *Intermodal Safety for Oil and Gas Transportation, Fraser Research Bulletin* (accessed 20 June 2020). https://www.fraserinstitute.org/sites/default/files/safety-first-intermodal-safety-for-oil-and-gas-transportation.pdf.
15. Furchtgott-Roth DE, Green K. *Intermodal Safety in the Transport of Oil. Studies in Energy Transportation*, 2013.
16. Green KP, Jackson T. *Safety in the Transportation of Oil and Gas: Pipelines or Rail?.* Vancouver, BC: Fraser Institute. 2015.
17. Belvederesi C, Thompson MS, Komers PE. Statistical analysis of environmental consequences of hazardous liquid pipeline accidents. *Heliyon.* 2018;4(11):e00901.
18. Accessed 20 June 2020, www.loc.gov.in.
19. Accessed 20 June 2020, https://indiacode.nic.in/handle/123456789/2401?view_type=browse&sam_handle=123456789/1362.
20. Accessed 20 June 2020, https://peso.gov.in/Petroleum_rule.aspx.
21. Accesses 20 June 2020, http://petroleum.nic.in/sites/default/files/PETROLEUM_RULES.pdf.
22. Accessed 20 June 2020, https://www.globalspec.com/learnmore/material_handling_packaging_equipment/material_handling_equipment/iso_containers.
23. *European Agreement Concerning the International Carriage of Dangerous Goods by Road (ADR 2017)*, United Nations Economic Commission for Europe (UNECE). 2016, (Accessed 20 June 2020), https://www.preventionweb.net/english/professional/policies/v.php?id=56430.
24. Martin S, Martin J, Lai P. International container design regulations and ISO standards: Are they fit for purpose?. *Maritime Policy & Management.* 2019;46(2):217–36.
25. Accessed 20 June 2020, http://www.isotank.in/.
26. Djukic LP, Rodgers DC, Herath MT. 3.16 Design, certification and field use of lightweight highly chemically resistant bulk liquid transport tanks. *Comprehensive Composite Materials* 2018;II(3):439–59.
27. Yildiz T. Design and analysis of a lightweight composite shipping container made of carbon fiber laminates. *Logistics.* 2019;3(3):18.
28. Accessed 20 June 2020, https://www.iso.org/obp/ui/#iso:std:iso:1496:-3:ed-4:en.
29. Accessed 20 June 2020, https://www.pressurevesselsindia.com/gallery/Presentation.pdf.
30. Accessed 20 June 2020, https://law.resource.org/pub/in/bis/S08/is.13187.1991.pdf.
31. Accessed 20 June 2020, https://www.meeberg.com/en/new-products/iso-tanks/20ft-diesel-tank/.
32. Accessed 20 June 2020, https://lavaengineering.in/t3-iso-tank-containers.php.

33. Fizal AN, Hossain MS, Alkarkhi AF, Oyekanmi AA, Hashim SR, Khalil NA, Zulkifli M, Yahaya AN. Assessment of the chemical hazard awareness of petrol tanker driver: A case study. *Heliyon.* 2019;5(8):e02368.

34. Belvederesi C, Thompson MS, Komers PE. Canada's federal database is inadequate for the assessment of environmental consequences of oil and gas pipeline failures. *Environmental Reviews.* 2017;25(4):415–22.

35. Accessed 20 June 2020, www.pennlease.com.

36. Accessed 20 June 2020, http://www.cybex.in/imports-data-india/iso-tank.aspx.

2 Mathematical Modelling and Reliability in Harvesting of Seafood for Food-Processing Industries

Rakesh Kumar and Navneet Rana
Shaheed Bhagat Singh State University
Guru Nanak College

CONTENTS

2.1 INTRODUCTION

In aquatic ecosystem, plankton is a general term that includes small plants and animals which freely float on the surface of water. They are very important part of marine ecology. Some of them are plants that are known as phytoplanktons. The phytoplanktons are usually green in colour and prepare their food by the process of photosynthesis. The growth of phytoplanktons depends mainly upon the nutrients available in the water and also on the temperature of water. The animal-like substances are known as zooplanktons. The zooplanktons depend upon the phytoplanktons for their food. Some of the zooplanktons eat other small zooplanktons. Fishes also play an important role in aquatic life. Fishes depend on zooplanktons and other small fishes for their food. Further, some of the zooplanktons like krill and jellyfish are harvested by humans for their food. It is important to maintain the stocks of these small zooplanktons in marine reserves. So, to understand the aquatic life, it is important to have analysis, which examines the effect of harvesting of fishes in phytoplanktons, zooplanktons and fish populations.

Many mathematical researchers have developed models which help understand the interaction of these microorganisms [1,2]. Freund et al. [3] explained about the extension of phytoplankton and zooplankton interaction model given by Truscott et al. [4]. This extension is always depending upon season, which forces the growth rate of phytoplanktons run by an oscillating temperature. The authors developed bistable behaviour of the model. They get a bloom and without bloom situation, the emphasis of time, find out switches between bloom and without bloom, and get that the blooms are interlinked with a fast temperature fluctuations upward and hypothesize on the role of blooms as trigger mechanisms. Edwards [5] formulated a two plankton population models with nutrients as well as detritus and analysed sensitivities to the model chaotic nature. The feeding policies are employed by zooplanktons in the sea. The authors in [6] studied the nutrient effect on the growth of phytoplanktons along with harvesting of fish population. They have examined biological and bionomical points of the model. Applying Pontryagin's principle, the authors have stressed upon the role of optimal policy of harvesting. The bifurcation analysis of the model has developed chaotic behaviour of the system. A food chain model with distinct kind of diseases in pests and gestation delay in natural enemy has been discussed by the authors in [7]. Per [8], fear plays an important role in the dynamical behaviour of three species in a defined food chain system, and the study examined it in a thorough way. The authors in [9] used the harvesting effect in the coexistence of prey-predator model and involved the exclusion of competitive predators. The authors in [10] studied the interaction between toxin-producing phytoplanktons and zooplanktons and their role in plankton ecology. In [11], the authors analysed the effect of delay in predation and with additional food in a fish–plankton dynamics. The effect of toxin on the plankton system is discussed by researchers in [12–15]. The authors in [16–18] discussed the optimal harvesting policy of phytoplankton–zooplankton system subject to the certain control constraints and impact of tax on harvesting policy. The researchers have given stress on the continuous harvesting of zooplanktons. They performed stability and global stability of the endemic equilibrium points and investigated the Hopf bifurcation analysis. Further, they find out transcritical kind of bifurcations in the defined system. They discussed that by imparting a tax according to the units of zooplankton, the optimal harvesting policy can be disposed of. Pontryagin's maximum principle has been applied to determine the problem with optimal harvest policy and is solved with constraints and state equations.

In this chapter, we study the dynamics of system with three populations namely phytoplanktons, zooplanktons and fish. The system is analysed with time delay, which arises when zooplanktons migrate in a horizontal or a vertical direction. The rest of the chapter is organised as follows: In the next section, the mathematical model is formulated. In Section 2.3, the positivity of all solutions and their boundedness is discussed. The possibility of various equilibriums and their stability with and without delay along with the Hopf bifurcation is discussed in Section 2.4. After that in Section 2.5, a numerical simulation is done to support the analytical results. Finally, the outcomes of this mathematical model are discussed as conclusions in the final section.

2.2 FORMATION OF MATHEMATICAL MODEL

The mathematical model involves interaction between three populations, namely, phytoplanktons, zooplanktons and fish. Let $P(t)$ be the density of phytoplanktons, $Z(t)$ be the density of zooplanktons and $F(t)$ be the density of fish population at t. The phytoplankton population has a growth rate r and a carrying capacity as K. The conversion rate of phytoplankton to zooplankton is α_1 and zooplankton to fish is β_1. Suppose τ is the delay taken in zooplankton predation and is arisen by the time taken by the zooplankton population in vertical and horizontal migration due to fish, which is the top predator. Here, E is the harvesting efforts of fish population. The model is defined as follows:

$$\left.\begin{aligned}
\frac{dP}{dt} &= r\left(P(t) - \frac{P^2(t)}{K}\right) - \frac{\alpha P(t)Z(t-\tau)}{\gamma_1 + P(t)}; \\
\frac{dZ}{dt} &= \frac{\alpha_1 P(t)Z(t-\tau)}{\gamma_1 + P(t)} - \delta_1 Z - \frac{\beta Z(t)F(t)}{\gamma_2 + Z(t)}; \\
\frac{dF}{dt} &= \frac{\beta_1 Z(t)F(t)}{\gamma_2 + Z(t)} - (\delta_2 + E)F(t).
\end{aligned}\right\}
\tag{2.1}$$

The various parameters involved are defined in Table 2.1.

Let $C\big([-\tau,0],R_+^3\big)$ be the Banach space, which is a collection of continuous functions defined on the set $[-\tau,0]$ into R_+^3, where $R_+^3 = \{(x_1,x_2,x_3): x_i > 0, i = 1,2,3\}$.

Then, $\phi_1(\theta), \phi_2(\theta), \phi_3(\theta) \in C\big([-\tau,0],R_+^3\big)$.

The initial condition for Eq. (2.1) is

TABLE 2.1
Parameters Description

Parameter	Description of Parameter
R	Growth rate of $P(t)$
K	Carrying capacity of $P(t)$
α	Capture rate of $Z(t)$ on $P(t)$
γ_1	Half saturation constant for $P(t)$
α_1	Conversion rate of $P(t)$ to $Z(t)$
δ_1	Natural mortality rate of $Z(t)$
β	Capture rate of $F(t)$ on $Z(t)$
γ_2	Half saturation constant for $Z(t)$
β_1	Conversion rate of $Z(t)$ to $F(t)$
δ_2	Natural mortality rate of $F(t)$
E	Harvesting efforts

$$P(\theta) = \phi_1(\theta), Z(\theta) = \phi_2(\theta),$$

$$F(\theta) = \phi_3(\theta); \phi_1(\theta) \geq 0, \phi_2(\theta) \geq 0, \phi_3(\theta) \geq 0, \theta \in [-\tau, 0]$$

$$\text{and } \phi_1(\theta) > 0, \phi_2(\theta), \phi_3(\theta) > 0.$$

(2.2)

2.3 POSITIVITY AND BOUNDEDNESS OF SOLUTIONS

In the present section, we give the conditions for the positive solution of the system (2.1) and also discuss the boundedness of the positive solutions.

Lemma 2.1

All the possible solutions of the mathematical system given in Eq. (2.1) subject to the initial conditions as defined in Eq. (2.2) are positive $\forall t \geq 0$.

Proof: Let the solution of Eq. (2.1) be $(P(t), Z(t), F(t))$ subject to initial conditions. We can rearrange the first equation of (2.1) and write it as

$$\frac{dP}{P} = \left\{ r\left(1 - \frac{P(t)}{K}\right) - \frac{\alpha Z(t-\tau)}{\gamma_1 + P(t)} \right\} dt,$$

$$\Rightarrow \frac{dP}{P} = \eta(P, Z) dt, \qquad \text{where } \eta(P, Z) = r\left(1 - \frac{P(t)}{K}\right) - \frac{\alpha Z(t-\tau)}{\gamma_1 + P(t)}.$$

Integrating from 0 to t, we achieve

$$\Rightarrow P(t) = \phi_1(0) \exp\left\{ \int_0^t \eta(P, Z) dt \right\} > 0 \quad \text{for } t \geq 0.$$

Consider the second equation of (2.1),

$$\frac{dZ}{dt} = \frac{\alpha_1 P(t) Z(t-\tau)}{\gamma_1 + P(t)} - \delta_1 Z - \frac{\beta Z(t) F(t)}{\gamma_2 + Z(t)},$$

$$\Rightarrow \frac{dZ}{dt} \geq -\delta_1 Z - \frac{\beta Z(t) F(t)}{\gamma_2 + Z(t)},$$

$$\Rightarrow \frac{dZ}{Z} \geq \left\{ -\delta_1 - \frac{\beta F(t)}{\gamma_2 + Z(t)} \right\} dt,$$

$$\Rightarrow \frac{dZ}{Z} \geq \left\{ \chi(Z, F) \right\} dt,$$

$$\text{where } \chi(Z, F) = -\delta_1 - \frac{\beta F(t)}{\gamma_2 + Z(t)}.$$

Integrating from 0 to t, we get

$$Z(t) \geq \phi_2(0) \exp\left\{\int_0^t \chi(Z,F)\,dt\right\} > 0 \text{ for } t \geq 0.$$

From the third equation of (2.1),

$$\frac{dF}{dt} = \frac{\beta_1 Z(t) F(t)}{\gamma_2 + Z(t)} - (\delta_2 + E)F(t),$$

$$\Rightarrow \frac{dF}{F} = \left\{\frac{\beta_1 Z(t)}{\gamma_2 + Z(t)} - (\delta_2 + E)\right\} dt,$$

$$\Rightarrow \frac{dF}{F} = \{\xi(Z,F)\} dt,$$

$$\text{where } \xi(Z,F) = \frac{\beta_1 Z(t)}{\gamma_2 + Z(t)} - (\delta_2 + E)$$

Integrating from 0 to t, we get

$$\Rightarrow F(t) = \phi_3(0) \exp\left\{\int_0^t \xi(Z,F)\,dt\right\} > 0 \text{ for } t \geq 0.$$

Thus, we get that all $P(t), Z(t), F(t) > 0$ for positive times.

Lemma 2.2

The solutions of the mathematical system defined in Eq. (2.1) are always uniformly bounded in the positive quadrant R_+^3.

Proof: Assume $(P(t), Z(t), F(t))$ is any solution of (2.1) with positive initial conditions defined in (2.2). From the first equation of (2.1),

$$\frac{dP}{dt} \leq r\left(1 - \frac{P(t)}{K}\right)P(t).$$

Using Comparison theorem given in [19],

$$\lim_{t \to \infty} \text{Sup } P(t) \leq K.$$

Define $\Theta(t) = \alpha_1 \beta_1 P + \alpha \beta_1 Z + \alpha \beta F$. We can have

$$\frac{d\Theta}{dt} = \alpha_1\beta_1\frac{dP}{dt} + \alpha\beta_1\frac{dZ}{dt} + \alpha\beta\frac{dF}{dt},$$

$$\Rightarrow \frac{d\Theta}{dt} = -\alpha_1\beta_1 rP - \alpha\beta_1\delta_1 Z - \alpha\beta\delta_2 F - \alpha_1\beta_1\frac{P^2 r}{K} + 2\alpha_1\beta_1 rP - \alpha\beta EF,$$

$$\Rightarrow \frac{d\Theta}{dt} \le -\Theta\psi + 2\alpha_1\beta_1 rP,$$

where $\psi = \min\{r, \delta_1, \delta_2\}$.

Then, $\dfrac{d\Theta}{dt} + \Theta\psi \le 2\alpha_1\beta_1 rP.$

By calculations, we get

$$\lim_{t\to\infty} \operatorname{Sup}\ \Theta(t) \le \frac{2\alpha_1\beta_1 rK}{\psi}. \tag{2.3}$$

Therefore, all values of $(P(t), Z(t), F(t))$ are bounded.

2.4 EXISTENCE OF EQUILIBRIA WITH STABILITY ANALYSES

Here, the existences of equilibrium points are given and follow with analyses for their stability. The distinct points are obtained as follows:

i. Null equilibrium $E_0(0,0,0)$, which covers the total disappearance of phyto-planktons, zooplanktons and fishes.
ii. The equilibrium $E_1(K,0,0)$, which covers the disappearance of zooplank-tons and fishes.
iii. The equilibrium $E_2(P', Z', 0)$, which covers the disappearance of fishes and without any delay, where

$$P' = \frac{\delta_1\gamma_1}{\alpha_1 - \delta_1} \text{ and } Z' = \frac{r\alpha_1\gamma_1}{\alpha}\left[\frac{K(\alpha_1 - \delta_1) - \delta_1\gamma_1}{K(\alpha_1 - \delta_1)^2}\right].$$

The equilibrium point $E_2(P', Z', 0)$ exists if the underwritten axioms are justified:

$$\alpha_1 > \delta_1 \text{ and } \alpha_1 > \delta_1 + \frac{\delta_1\gamma_1}{K},$$

which gives

$$\alpha_1 > \max\left\{\delta_1, \delta_1 + \frac{\delta_1\gamma_1}{K}\right\}.$$

This gives $\alpha_1 > \delta_1 + \dfrac{\delta_1 \gamma_1}{K}$.

   iv. The coexisting equilibrium $E_3(P^*, Z^*, F^*)$ of Eq. (2.1) by solving it for $\tau = 0$
     as

$$-\frac{\alpha P(t)Z(t)}{\gamma_1 + P(t)} + r\left(P(t) - \frac{P^2(t)}{K}\right) = 0;$$

$$-\delta_1 Z - \frac{\beta Z(t)F(t)}{\gamma_2 + Z(t)} + \frac{\alpha_1 P(t)Z(t)}{\gamma_1 + P(t)} = 0; \qquad (2.4)$$

$$\frac{\beta_1 Z(t)F(t)}{\gamma_2 + Z(t)} - (\delta_2 + E)F(t) = 0.$$

Solving system of Eq. (2.4), we get

$$Z^* = \frac{\gamma_2(\delta_2 + E)}{\beta_1 - \delta_2 - E}. \qquad (2.5)$$

Also P^* is a value of quadratic equation given as

$$P^{*2} + (\gamma_1 - K)P^* + \frac{K\alpha\gamma_2(\delta_2 + E)}{r(\beta_1 - \delta_2 - E)} - K\gamma_1 = 0, \qquad (2.6)$$

$$F^* = \frac{\gamma_2 + Z^*}{\beta}\left(\frac{\alpha_1 P^*}{\gamma_1 + P^*} - \delta_1\right). \qquad (2.7)$$

For the existence of the coexisting equilibrium point, the following conditions should
be satisfied:

$$\beta_1 > \delta_2 + E \; ; \; \gamma_1 > K \; ;(\alpha\gamma_2 + \gamma_1 r)(\delta_2 + E) < \gamma_1 r \beta_1.$$

To study the behaviour of solution near the equilibriums, we compute the Jacobian
matrix at any point (P, Z, F) as

$$J_E = \begin{pmatrix} A_{11} & A_{12} & 0 \\ A_{21} & A_{22} & A_{23} \\ 0 & A_{32} & A_{33} \end{pmatrix},$$

where

$$A_{11} = -\frac{\alpha Z \gamma_1}{(\gamma_1 + P)^2} + r\left(1 - \frac{2P}{K}\right);$$

$$A_{12} = -\frac{\alpha P}{\gamma_1 + P};$$

$$A_{21} = \frac{\alpha_1 \gamma_1 Z}{(\gamma_1 + P)^2};$$

$$A_{22} = \frac{\alpha_1 P}{\gamma_1 + P} - \delta_1 - \frac{\beta F \gamma_2}{(\gamma_2 + Z)^2}; A_{23} = -\frac{\beta Z}{\gamma_2 + Z};$$

$$A_{32} = \frac{\beta_1 F \gamma_2}{(\gamma_2 + Z)^2}; A_{33} = \frac{\beta_1 Z}{\gamma_2 + Z} - (\delta_2 + E).$$

2.4.1 STABILITY ANALYSIS

Next, we will find the stability conditions of equilibrium points. By linearizing (1) in the neighbourhood of equilibrium, we can obtain the criterion for local stability of the distinct equilibria in the subsequent result:

Theorem 2.1

i. The null equilibrium $E_0(0,0,0)$ is not stable.

ii. The equilibrium $E_1(K,0,0)$ is L.A.S. if $\delta_1 > \frac{\alpha_1 K}{\gamma_1 + K}$.

iii. The equilibrium point $E_2(P',Z',0)$ is L.A.S. for $\tau = 0$ if

$$\delta_2 + E > \frac{\beta_1 Z'}{\gamma_2 + Z'} \text{ and } r\left(1 - \frac{2P'}{K}\right) < \frac{\alpha Z' \gamma_1}{(\gamma_1 + P)^2}.$$

Proof: The Jacobian matrix at null equilibrium point $E_0(0,0,0)$ is

$$J_{E_0} = \begin{pmatrix} r & 0 & 0 \\ 0 & -\delta_1 & 0 \\ 0 & 0 & -(\delta_2 + E) \end{pmatrix}.$$

The eigenvalues of the Jacobian matrix J_{E_0} are $r, -\delta_1, -(\delta_2 + E)$. The real part of the first eigenvalue is positive. Hence, the equilibrium $E_0(0,0,0)$ is not a stable point.

i. The Jacobian matrix at the equilibrium point $E_1(K,0,0)$ is given by

$$J_{E_1} = \begin{pmatrix} -r & -\dfrac{\alpha K}{\gamma_1 + K} & 0 \\ 0 & \dfrac{\alpha_1 K}{\gamma_1 + K} - \delta_1 & 0 \\ 0 & 0 & -(\delta_2 + E) \end{pmatrix}.$$

The eigenvalues of the Jacobian matrix J_{E_1} are $\dfrac{\alpha_1 K}{\gamma_1 + K} - \delta_1, -(\delta_2 + E), -r$.

Hence, $E_1(K,0,0)$ is L.A.S. if $\delta_1 > \dfrac{\alpha_1 K}{\gamma_1 + K}$.

ii. The Jacobian matrix at $E_2(P', Z', 0)$ for $\tau = 0$ is

$$J_{E_2} = \begin{pmatrix} p_1 & -p_2 & 0 \\ p_3 & 0 & -p_4 \\ 0 & 0 & p_5 \end{pmatrix},$$

where

$$p_1 = r\left(1 - \frac{2P'}{K}\right) - \frac{\alpha Z' \gamma_1}{(\gamma_1 + P')^2}; -p_2 = -\frac{\alpha P'}{\gamma_1 + P'}; p_3 = \frac{\alpha_1 \gamma_1 Z'}{(\gamma_1 + P')^2};$$

$$-p_4 = -\frac{\beta Z'}{\gamma_2 + Z'}; p_5 = \frac{\beta_1 Z'}{\gamma_2 + Z'} - (\delta_2 + E).$$

The characteristic equation of J_{E_2} is $(p_5 - \lambda)(\lambda^2 - p_1\lambda + p_2 p_3) = 0$. This characteristic equation has real parts with a negative sign if $p_5 < 0$ and $p_1 < 0$

i.e. $\dfrac{\beta_1 Z'}{\gamma_2 + Z'} - (\delta_2 + E) < 0 \rightarrow \dfrac{\beta_1 Z'}{\gamma_2 + Z'} < (\delta_2 + E)$ or $(\delta_2 + E) > \dfrac{\beta_1 Z'}{\gamma_2 + Z'}$

and $r\left(1 - \dfrac{2P'}{K}\right) - \dfrac{\alpha Z' \gamma_1}{(\gamma_1 + P')^2} < 0 \rightarrow r\left(1 - \dfrac{2P'}{K}\right) < \dfrac{\alpha Z' \gamma_1}{(\gamma_1 + P')^2}$.

Theorem 2.2

The coexisting equilibrium point $E_3(P^*, Z^*, F^*)$ is locally asymptotically stable if the following conditions are satisfied:

Proof: The Jacobian matrix at $E_3(P^*, Z^*, F^*)$ is

$$J_{E_3} = \begin{pmatrix} r\left(1 - \dfrac{2P^*}{K}\right) - \dfrac{\gamma_1 \alpha Z^*(t-\tau)}{\left(\gamma_1 + P^*(t)\right)^2} & -\dfrac{\alpha P^* e^{-\lambda\tau}}{\gamma_1 + P^*(t)} & 0 \\[2ex] \dfrac{\gamma_1 \alpha_1 Z^*(t-\tau)}{\left(\gamma_1 + P^*(t)\right)^2} & \dfrac{\alpha_1 P^* e^{-\lambda\tau}}{\gamma_1 + P^*(t)} - \delta_1 - \dfrac{\gamma_2 \beta F^*}{(\gamma_2 + Z^*(t))^2} & -\dfrac{\beta Z^*}{\gamma_2 + Z^*(t)} \\[2ex] 0 & \dfrac{\gamma_2 \beta_1 F^*}{(\gamma_2 + Z^*(t))^2} & \dfrac{\beta_1 Z^*}{\gamma_2 + Z^*(t)} - (\delta_2 + E) \end{pmatrix}.$$

The characteristic equation of J_{E_3} is

$$(\lambda^3 + A\lambda^2 + B\lambda + C) + e^{-\lambda\tau}(D\lambda^2 + E\lambda + F) = 0, \qquad (2.8)$$

where

$$A = -\left[r\left(1 - \frac{2P^*}{K}\right) - \frac{\gamma_1 \alpha Z^*(t-\tau)}{\left(\gamma_1 + P^*\right)^2} - \delta_1 + \frac{\beta_1 Z^*}{\left(\gamma_2 + Z^*\right)} - (\delta_2 + E) - \frac{\gamma_2 \beta F^*}{\left(\gamma_2 + Z^*\right)^2} \right],$$

$$B = -\left[r\left(1 - \frac{2P^*}{K}\right)\delta_1 - r\left(1 - \frac{2P^*}{K}\right)\frac{\beta_1 Z^*}{\gamma_2 + Z^*} + r\left(1 - \frac{2P^*}{K}\right)(\delta_2 + E) + r\left(1 - \frac{2P^*}{K}\right)\frac{\gamma_2 \beta F^*}{(\gamma_2 + Z^*)^2} \right.$$
$$- \frac{\gamma_1 \alpha Z^*(t-\tau)\delta_1}{(\gamma_1 + P^*)^2} + \frac{\gamma_1 \alpha Z^*(t-\tau)\beta_1 Z^*}{(\gamma_1 + P^*)^2(\gamma_2 + Z^*)} - \frac{\gamma_1 \alpha Z^*(t-\tau)(\delta_2 + E)}{(\gamma_1 + P^*)^2} - \frac{\gamma_1 \alpha Z^*(t-\tau)\gamma_2 \beta F^*}{(\gamma_1 + P^*)^2(\gamma_2 + Z^*)^2}$$
$$\left. + \frac{\beta_1 Z^* \delta_1}{(\gamma_2 + Z^*)} - \delta_1(\delta_2 + E) - \frac{\gamma_2 \beta F^*(\delta_2 + E)}{(\gamma_2 + Z^*)^2} \right],$$

$$C = -\left[-r\left(1 - \frac{2P^*}{K}\right)\frac{\delta_1 \beta_1 Z^*}{\gamma_2 + Z^*} + r\left(1 - \frac{2P^*}{K}\right)\delta_1(\delta_2 + E) + r\left(1 - \frac{2P^*}{K}\right)\frac{\gamma_2 \beta F^*(\delta_2 + E)}{(\gamma_2 + Z^*)^2} \right.$$
$$\left. + \frac{\gamma_1 \alpha Z^*(t-\tau)\delta_1 \beta_1 Z^*}{(\gamma_1 + P^*)^2(\gamma_2 + Z^*)} - \frac{\gamma_1 \alpha Z^*(t-\tau)\delta_1(\delta_2 + E)}{(\gamma_1 + P^*)^2} - \frac{\gamma_1 \alpha Z^*(t-\tau)\gamma_2 \beta F^*(\delta_2 + E)}{(\gamma_1 + P^*)^2(\gamma_2 + Z^*)^2} \right],$$

$$D = -\frac{\alpha_1 P^*}{\gamma_1 + P^*},$$

$$E = \left[r\left(1 - \frac{2P^*}{K}\right)\frac{\alpha_1 P^*}{\gamma_1 + P^*} - \frac{\alpha_1 P^*(\delta_2 + E)}{\gamma_1 + P^*} + \frac{\alpha_1 \beta_1 P^* Z^*}{(\gamma_1 + P^*)(\gamma_2 + Z^*)} \right],$$

$$F = \left[-r\left(1 - \frac{2P^*}{K}\right)\frac{\alpha_1 \beta_1 P^* Z^*}{(\gamma_1 + P^*)(\gamma_2 + Z^*)} + r\left(1 - \frac{2P^*}{K}\right)\frac{\alpha_1 P^*(\delta_2 + E)}{\gamma_1 + P^*} \right].$$

For $\tau = 0$, Eq. (2.8) becomes

$$\lambda^3 + (A+D)\lambda^2 + (B+E)\lambda + (C+F) = 0. \tag{2.9}$$

The coexisting equilibrium is L.A.S. (locally asymptotically stable) if the coefficients of (2.8), $(A+D)$, $(B+E)$ and $(C+F)$, satisfy the Routh–Hurwitz stability criterion [20], i.e.

$$S_1 : \text{(i) } A+D > 0, \text{ (ii) } B+E > 0, \text{ (iii) } C+F > 0,$$
$$\text{(iv) } (A+D)(B+E) - (C+F) > 0. \tag{2.10}$$

2.4.2 HOPF BIFURCATION

To analyse the Hopf bifurcation, we first state the lemma as follows:

Lemma 2.3

i. The equilibrium point $E_3\left(P^*, Z^*, F^*\right)$ of (2.1) is absolutely stable if the point $E_3(P^*, Z^*, F^*)$ of model (2.1) is L.A.S. and the transcendental characteristic Eq. (2.8) will have real roots for $\tau > 0$ [21].
ii. The equilibrium point $E_3(P^*, Z^*, F^*)$ of (2.1) is conditionally stable if all values of (2.8) have real parts with a negative sign (for zero τ), and $\exists \tau > 0$ so that the Eq. (2.8) has an imaginary eigenvalues of the form $\pm i\omega$.

The following theorem may also be stated:

Theorem 2.3

The equilibrium point $E_3(P^*, Z^*, F^*)$ is conditionally stable if (S_1) hold good for model (2.1) [22].

Let τ be the bifurcation parameter. Let for any $\tau > 0$, $\lambda = i\omega$ for any $\omega > 0$ be a root of (2.8). Put $\lambda = i\omega$ into Eq. (2.8) and find the real and imaginary parts of characteristic Eq. (2.8),

$$E\omega \sin(\omega\tau) + (-D\omega^2 + F)\cos(\omega\tau) = A\omega^2 - C, \tag{2.11}$$

$$E\omega \cos(\omega\tau) + (-F + D\omega^2)\sin(\omega\tau) = \omega^3 - B\omega. \tag{2.12}$$

Solving Eqs. (2.11) and (2.12), one can easily get after simple computations

$$\Sigma(s) = s^3 + \Omega_1 s^2 + \Omega_2 s + \Omega_3 = 0, \tag{2.13}$$

$$\text{where} \begin{cases} s = \omega^2, \text{ and} \\ \Omega_1 = A^2 - 2B - D^2, \\ \Omega_2 = B^2 - 2AC + 2FD - E^2, \\ \Omega_3 = C^2 - F^2 \end{cases} \tag{2.14}$$

If the one root of equation is non-negative, then Eq. (2.8) will have complex roots. So, if s_1 is the positive value of (2.13), then the complex conjugate pair of eigenvalues is given by $\omega = \pm\sqrt{s_1}$. Thus, the Hopf bifurcation might exist in the system (2.1). The next theorem explains the possibility of the existence of a positive real root of (2.13):

Theorem 2.4

Equation (2.13), $\Sigma(s) = s^3 + \Omega_1 s^2 + \Omega_2 s + \Omega_3 = 0$, has a minimum one non-negative root if either of the following conditions are justified [23]:

i. $\Omega_3 < 0$,

ii. $\Omega_3 \geq 0$, $\Omega_1^2 - 3\Omega_2 > 0$, and $s_c > 0$ of $\Sigma(s)$ exists with $\Sigma(s_c) \leq 0$.

Thus, the model system (2.1) has purely complex eigenvalues if (i) these conditions are justified or (ii) all conditions of Lemma 2.3 are verified. Also (2.13) may have at most three non-negative real roots, $s_i > 0$; $i = 1, 2, 3$, and there is the possibility that it may have three complex eigenvalues, $\lambda_i = i\omega_i = \pm i\sqrt{s_i}, i = 1, 2, 3$.

Let us now find the values of the time delay parameter τ_i corresponding to values of ω_i, and by applying ω_i into Eqs. (2.11) and (2.12) for the real and imaginary parts of the characteristic Eq. (2.8). Solving for $\sin(\omega_i\tau)$ and $\cos(\omega_i\tau)$, one can easily obtain the following:

$$\cos(\omega_i\tau) = \frac{(\omega_i^3 - B\omega_i)E\omega_i - (A\omega_i^2 - C)(D\omega_i^2 - F)}{E^2\omega_i^2 + (F - D\omega_i^2)^2}, \tag{2.15}$$

$$\sin(\omega_i\tau) = \frac{(A\omega_i^2 - C)E\omega_i - (\omega_i^3 - B\omega_i)(F - D\omega_i^2)}{E^2\omega_i^2 + (F - D\omega_i^2)^2}. \tag{2.16}$$

The thresh value of τ at which the system exhibits stability switch, i.e. values of (2.18) at which complex roots occur, can be obtained by using Eq. (2.15) and get that

$$\tau_i^{(j)} = \frac{1}{\omega_i}\cos^{-1}\left(\frac{(\omega_i^3 - B\omega_i)E\omega_i - (A\omega_i^2 - C)(D\omega_i^2 - F)}{E^2\omega_i^2 + (F - D\omega_i^2)^2}\right) + \frac{2j\pi}{\omega_i}, \quad (2.17)$$

where $i = 0,1,2$ and $j = 0,1,2,3,\ldots$ The least value of τ_0 at which the purely imaginary eigenvalues of the form $\lambda_0 = \pm i\omega_0$ occur is therefore given as below:

$$\tau_0 = \min_{0\le i\le2, j\ge0} \tau_i^{(j)}, \tau_i^{(j)} > 0.$$

Now, let us find the condition of the Hopf bifurcation [19] for model (2.1) in the neighbourhood of the interior point $E_3(P^*, Z^*, F^*)$. Taking τ as a parameter of bifurcation and suppose that $\lambda = \zeta + i\omega$ for $\tau > 0$ is a value of characteristic Eq. (2.8), here $\omega > 0$ is in the real domain. Putting $\lambda = \zeta + i\omega$ into (2.8), and obtained as follows:

$$\zeta^3 - 3\zeta\omega^2 + A(\zeta^2 - \omega^2) + B\zeta + C +$$

$$\left[\{D(\zeta^2 - \omega^2) + E\zeta + F\}\cos(\omega\tau) + (2D\zeta\omega + E\omega)\sin(\omega\tau)\right]e^{-\zeta\tau} = 0, \quad (2.18)$$

and

$$\left(-\omega^3 + 3\zeta^2\omega + 2A\zeta\omega + B\omega\right) +$$

$$\left[-\{D(\zeta^2 - \omega^2) + E\zeta + F\}\sin(\omega\tau) + (2D\zeta\omega + E\omega)\cos(\omega\tau)\right]e^{-\zeta\tau} = 0. \quad (2.19)$$

Differentiating Eqs. (2.18) and (2.19) w.r.t. τ and letting $\tau = \hat{\tau}, \omega = \hat{\omega}$, and $\zeta = 0$, the obtained expressions are as below:

$$A_1\left[\frac{d\zeta}{d\tau}\right]_{\tau=\hat{\tau}} - A_2\left[\frac{d\omega}{d\tau}\right]_{\tau=\hat{\tau}} = B_1, \quad (2.20)$$

$$A_2\left[\frac{d\zeta}{d\tau}\right]_{\tau=\hat{\tau}} + A_1\left[\frac{d\omega}{d\tau}\right]_{\tau=\hat{\tau}} = B_2, \quad (2.21)$$

where

$$\begin{cases} A_1 = -3\hat{\omega}^2 + B + \{E + \tau(D\omega^2 - F)\}\cos(\omega\tau) + \omega(2D - E\tau)\sin(\omega\tau), \\ A_2 = 2A\omega - \{E + \tau(D\omega^2 - F)\}\sin(\omega\tau) + +\omega(2D - E\tau)\cos(\omega\tau), \\ B_1 = \omega(F - D\omega^2)\sin(\omega\tau) - E\omega^2\cos(\omega\tau), \\ B_2 = \omega(F - D\omega^2)\cos(\omega\tau) + E\omega^2\sin(\omega\tau). \end{cases} \quad (2.22)$$

Simplify Eqs. (2.20) and (2.21), and at $\tau = \hat{\tau} = \tau_0$, $\omega = \hat{\omega}^2 = \omega_0{}^2$,

$$\left[\frac{d\zeta}{d\tau}\right]_{\tau=\tau_0} = \frac{\hat{\omega}^2}{A_1{}^2 + A_2{}^2}\left[\frac{d\Sigma}{ds}\right]_{\omega=\omega_0{}^2} \neq 0. \tag{2.23}$$

Hence, the values of the transcendental Eq. (2.8) cross the vertical axis as the bifurcation parameter τ crosses over the threshold value.

Hence, at $\tau = \tau_0$, and is the smallest non-negative value of τ presented by Eq. (2.17), the conditions for the Hopf bifurcation are justified.

Theorem 2.5

Suppose that $E_3(P^*, Z^*, F^*)$ exist and the conditions in S_1: (i) $A + D > 0$, (ii) $B + E > 0$, (iii) $C + F > 0$, (iv) $(A + D)(B + E) - (C + F) > 0$ are justified for model (1), and then the necessary and sufficient conditions for equilibrium point $E_3(P^*, Z^*, F^*)$ to be L.A.S. with time delay τ are [24]

i. $E_3(P^*, Z^*, F^*)$ is L.A.S. for $\tau \in [0, \tau_0)$.
ii. $E_3(P^*, Z^*, F^*)$ bifurcates into periodic orbits, i.e. becomes unstable for $\tau \geq \tau_0$.
iii. model (2.1) meets with the Hopf bifurcation at a threshold value τ_0 around $E_3(P^*, Z^*, F^*)$, where

$$\tau_0{}^* = \frac{1}{\omega_0}\cos^{-1}\left(\frac{(\omega_0{}^3 - B\omega_0)E\omega_0 - (A\omega_0{}^2 - C)(D\omega_0{}^2 - F)}{E^2\omega_0{}^2 + (F - D\omega_0{}^2)^2}\right). \tag{2.24}$$

2.5   NUMERICAL SIMULATIONS

The present section deals with the dynamical behaviour of the system by numerical computations. Let us take a set of values for distinct parametric for the system (2.1) as given below:

$$\left.\begin{aligned}
\frac{dP}{dt} &= (0.35)\left(P(t) - \frac{P(t)^2}{10}\right) - \frac{(1.5)P(t)Z(t-\tau)}{10 + P(t)}; \\
\frac{dZ}{dt} &= \frac{(1.2)P(t)Z(t-\tau)}{10 + P(t)} - (0.1)Z - \frac{(1.5)Z(t)F(t)}{20 + Z(t)}; \\
\frac{dF}{dt} &= \frac{(1.12)Z(t)F(t)}{20 + Z(t)} - (0.07 + 0.001)F(t).
\end{aligned}\right\} \tag{2.25}$$

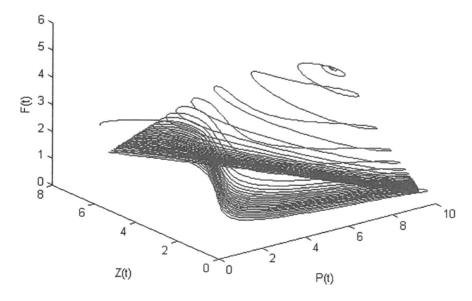

FIGURE 2.1   The stable solution curves of the system at $\tau = 3.6$

By using $P(0) = 1.3$, $Z(0) = 7.1$ and $F(0) = 2.3$ as initial values in (2.25), the system has been computed with the help of Matlab package DDE23 and showed a stable dynamical behaviour for $\tau = 3.6 < 3.7 = \tau_0^*$, and is clearly depicted in Figure 2.1. The system (2.25) produced periodic oscillations of the solution curves of all the populations for a critical value $\tau_0^* = 3.7$,

which is shown in Figure 2.2. Table 2.2 shows the variation in the density of phytoplanktons, zooplanktons and fish population.

In Table 2.2, we see that with a little increase in the growth rate of population $P(t)$, there is an increase in the phytoplankton population density $P(t)$, which consequently helps to increase the fish density $F(t)$ due to capture rate.

Further, when the efforts of harvesting fish population density have been increased from 0.1% to 5%, the density of fish population goes on decreasing and converges to zero. This is shown in Figure 2.2a and b. It shows that the optimal harvesting efforts policies are to be given by the government, and all the governments (central/state/UT) provide these policies from time to time through their corresponding departments for the enhancements of the seafood (fishes) in the food-processing industries.

2.6   RESULTS

In this chapter, we have studied a three-dimensional model (time delayed) to learn about the interaction between phytoplankton, zooplankton and fish populations. By taking the predation time of zooplankton population τ as the bifurcation parameter, we observed that it has affected the stability properties of non-negative point $E_3(P^*, Z^*, F^*)$ of model (2.1) at a critical value $\tau_0^* = 3.7$. When there is an

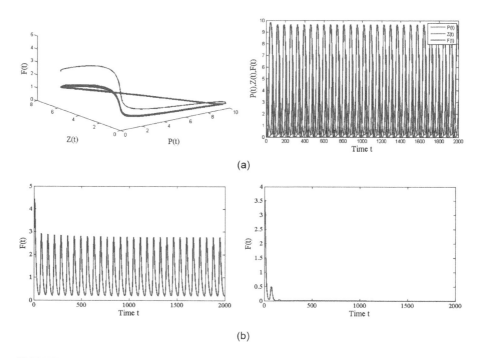

FIGURE 2.2   Periodic behaviour is shown for the system (2.25) with a critical value $\tau_0^* = 3.7$. (a) Density of fish population at $E = 0.001$. (b) Density of fish population at $E = 0.05$.

TABLE 2.2
Variation of Plankton Density with
Growth Rate of Phytoplankton

Growth Rate (r)	P(t)	Z(t)	F(t)
0.35	4.281	1.518	1.033
0.5	7.559	1.372	5.826
0.8	8.576	1.367	6.461
1.2	9.077	1.369	6.739
1.5	9.267	1.373	6.861

enhancement in the delay parameter beyond $\tau_0^* \geq 3.7$, the periodic oscillation arises due to the Hopf bifurcation in (2.1).

We all know about this fact that the fishing industry is the main industry of our country and is giving employment to 14.5 million people. This industry covers the activities related with taking of fishes from the water, their culturing, processing and preserving of different kinds of fishes, their storage, transportation, the selling of fishes or fish-related products, and their marketing. Further, as we all know about this fact that the central government has always extended financial help to fishermen with the help of various policies being run through the state governments as well as central fishery department for the motorization and mobilization of traditional fishermen. The various governments have always extended their support for constructing fishing harbours, setting up fish landing centres, giving rebates on high speed diesel (HSD) oil, and establishing fish marketing centres. The governments in their 5-year plans always provide assistance to increase the production of fisheries and helped to strengthen the infrastructure to provide fish in prime condition to the consumers and to various fish processing industries. Also, the government supports harvesting efforts to enhance the harvesting of seafood (fishes) in the food-processing industries. This has been done through machines. These machines use completely innovative types of gear which may separate the **fish** species or shellfish under the water with the help of pumps. In the same context, if we observe mathematical model (2.1), we noticed that the growth rate of $P(t)$ is also responsible for the increase in fish population. It has showed that the rise in phytoplankton density played an exemplary role for the upliftment of fishing industries. The harvesting efforts E also played an important role in this process. As we increase the value of harvesting efforts E, the fish population starts decreasing. It means that the extensive rise in harvesting proved that further no fish is available for harvesting. So, by controlling harvesting efforts, making an optimal policy of harvesting, and by increasing growth rate of phytoplanktons, the fish population has been maintained at an optimal level for the fishing process. It will always be a very useful aspect in the fishing industry and for the enrichment of the seafood (fishes) for the food-processing units/industries of the country.

REFERENCES

1. Nakaoka S, Takeuchi Y. Competition in chemostat-type equations with two habitats. *Mathematical Biosciences.* 2006;201(1–2):157–171.
2. Morozov AY. Emergence of Holling type III zooplankton functional response: Bringing together field evidence and mathematical modelling. *Journal of Theoretical Biology.* 2010;265(1):45–54.
3. Freund JA, Mieruch S, Scholze B, Wiltshire K, Feudel U. Bloom dynamics in a seasonally forced phytoplankton–zooplankton model: Trigger mechanisms and timing effects. *Ecological Complexity.* 2006;3(2):129–139.
4. Truscott JE, Brindley J. Ocean plankton populations as excitable media. *Bulletin of Mathematical Biology.* 1994;56(5):981–98.
5. Edwards AM. Adding detritus to a nutrient–phytoplankton–zooplankton model: A dynamical-systems approach. *Journal of Plankton Research.* 2001;23(4):389–413.
6. Dubey B, Patra A, Upadhyay RK. Dynamics of phytoplankton, zooplankton and fishery resource model. *Applications and Applied Mathematics: An International Journal (AAM).* 2014;9(1):14.
7. Kumar V, Dhar J, Bhatti HS, Singh H. Plant-pest-natural enemy dynamics with disease in pest and gestation delay for natural enemy. *Journal of Mathematics and Computer Science.* 2017;7(5):948–965.
8. Kumar V, Kumari N. Controlling chaos in three species food chain model with fear effect. *AIMS Mathematics.* 2020;5(2):828–842.

9. Pei Y, Lv Y, Li C. Evolutionary consequences of harvesting for a two-zooplankton one-phytoplankton system. *Applied Mathematical Modelling*. 2012;36(4):1752–65.

10. Sharma A, Sharma AK, Agnihotri K. Analysis of a toxin producing phytoplankton–zooplankton interaction with Holling IV type scheme and time delay. *Nonlinear Dynamics*. 2015;81(1):13–25.

11. Kaur RP, Sharma A, Sharma AK. Dynamical complexity in a delayed Plankton-Fish model with alternative food for predators. *Numerical Algebra, Control & Optimization*. 2021.

12. Chattopadhayay J, Sarkar RR, Mandal S. Toxin-producing plankton may act as a biological control for planktonic blooms—field study and mathematical modelling. *Journal of Theoretical Biology*. 2002;215(3):333–44.

13. Zhao Q, Liu S, Niu X. Effect of water temperature on the dynamic behavior of phytoplankton–zooplankton model. *Applied Mathematics and Computation*. 2020;378:125211.

14. Chattopadhyay J, Sarkar RR, El Abdllaoui A. A delay differential equation model on harmful algal blooms in the presence of toxic substances. *Mathematical Medicine and Biology: A Journal of the IMA*. 2002;19(2):137–61.

15. Saha T, Bandyopadhyay M. Dynamical analysis of toxin producing phytoplankton–zooplankton interactions. *Nonlinear Analysis: Real World Applications*. 2009;10(1):314–32.

16. Rehim M, Imran M. Dynamical analysis of a delay model of phytoplankton–zooplankton interaction. *Applied Mathematical Modelling*. 2012;36(2):638–47.

17. Jang S, Baglama J, Wu L. Dynamics of phytoplankton–zooplankton systems with toxin producing phytoplankton. *Applied Mathematics and Computation*. 2014;227:717–40.

18. Wang Y, Wang H. Stability and selective harvesting of a phytoplankton–zooplankton system. *Journal of Applied Mathematics*. 2014;2014.

19. Birkhoff G, Rota G. *Ordinary Differential Equations*, Ginn, Boston, MA. 1989.

20. Luenberger DG. *Introduction to Dynamic Systems: Theory, Models, and Applications*. 1979.

21. Kumar R, Sharma AK, Agnihotri K. Bifurcation analysis of a nonlinear diffusion model: Effect of evaluation period for the diffusion of a technology. *Arab Journal of Mathematical Sciences*. 2019;25(2):189–213.

22. Kumar R. Mathematical modeling of a delayed innovation diffusion model with media coverage in adoption of an innovation. *Systems Reliability Engineering*. 2021:153–72.

23. Kumar R. The dynamics of a continuous innovation diffusion model with advertisements as well as interpersonal communications. In *Applications of Advanced Optimization Techniques in Industrial Engineering*, 2022 (pp. 79–100). CRC Press.

24. Kumar R, Sharma AK, Agnihotri K. Hopf bifurcation analysis in a multiple delayed innovation diffusion model with Holling II functional response. *Mathematical Methods in the Applied Sciences*. 2020;43(4):2056–75.

3 Petri Net-Based Approach for Performance Modeling and Availability Analysis of a Butter Oil Processing Plant

Ankur Bahl
Lovely Professional University

Anish Sachdeva
Dr B.R. Ambedkar National Institute of Technology

Munish Mehta
Lovely Professional University

CONTENTS

3.1   INTRODUCTION

The operation and maintenance of the butter oil processing plant involves the study of complex subsystems whose failures of various units lead to the plant unavailability. During the operation of the plant, the failure of subsystems may occur due to many reasons such as wear and tear of the equipment or some faults related to sensing devices. Also, faulty operating procedures adopted by equipment operators may sometimes lead to unavoidable failures. Failure is an unavoidable phenomenon. The availability of the plant can be enhanced by increasing the redundant parts or by increasing the reliability of the components. The increase in availability is achieved if the components are more reliable and more effective maintenance measures are being adopted. As availability is a function of both reliability and maintainability, they are used to assess the plant performance. In broader terms, the availability of the system is measured as the proportion of uptime to the total time the system is in service. Therefore,

$$\text{Availability} = \frac{\text{Up Time}}{\text{Up Time} + \text{Down Time}}$$

Based upon this equation, the availability of the complex systems can be calculated. Various tools and techniques have been discussed by various researchers for the availability analysis of complex systems in the literature. Mehta et al. [1] carried out the availability analysis of the butter oil production system using the Supplementary Variable Technique. The effect of failure and repair rate (FRR) of various units on the system availability was discussed. Zang et al. [2] used six sigma and Gauss–Legendre quadrature formula for carrying out the reliability analysis of a complex system. Byun et al. [3] presented a matrix-based system reliability method to identify the intricate dependence between the various component failures for reliability analysis.

Aggarwal et al. [4] investigated the reliability, availability and maintainability (RAM) analysis of tunnel boring machines using Markov chains. The effect of failure and repair rates of various components of the machine on the RAM analysis was analyzed. Pandey et al. [5] carried out the reliability analysis of dragline using failure mode and effect analysis and Bathtub curve. Gupta et al. [6] presented the reliability analysis of the butter oil manufacturing plant using Markov chains considering the constant FRRs of the components. Garg et al. [7] used a Supplementary Variable Technique for carrying out the availability optimization under steady state for yarn plant. Khorshidi et al. [8] studied the reliability analysis of complex large systems using Failure Mode Effect and Criticality Analysis technique. Murthy et al. [9] performed the reliability analysis using a Markov model of the phase measurement unit considering the transient system states. Bourouni [10] carried out the availability analysis of an reverse osmosis (RO) using both the Fault Tree and the Reliability Block Diagram considering economic optimization as one of the main objectives. Imakhlaf et al. [11] discussed the reliability analysis of the non-coherent systems using binary decision diagrams. Choi and Chang [12] proposed the Fault Tree model to analyze the reliability and availability of seabed storage tanks. A four-step procedure of data collection and modeling the system for the reliability assessment was proposed. Modgil et al. [13] analyzed the performance of a shoe upper manufacturing plant based upon the Markov approach. Ahmed et al. [14] presented the availability analysis of gas sweetening plant. Though the Reliability Block Diagram (RBD) and Fault Trees have been used by some researchers for the RAM analysis, it is difficult to build RBD and Fault Trees for the complex systems.

Many researches have used the Markov chains and Supplementary Variable Technique for the reliability analysis of the complex systems. These techniques are helpful in the calculation of long-run availability, but they are difficult to formulate. Moreover, they involve complex differential equations to calculate the availability of the system that needs more computational efforts. It is difficult to form state transition diagrams for complex systems. These traditional tools and techniques do not address problems like complex parallelism, process dependency, and resource constraints. These problems are well addressed by Petri nets, which is an effective modeling tool to represent the complex systems. Petri net being a graphical tool is used as communication visual aid and being a mathematical tool helps to study the dynamic behavior of the system by setting up the governing equation to the model.

Many authors have discussed the various applications of Petri nets in different fields. Bahl et al. [15] discussed the use of Petri nets in the availability analysis of a distillery plant. Sachdeva et al. [16] presented the application of stochastic Petri nets to study the behavior of the feeding system of a paper plant. Patel and Joshi [17] carried out the analysis of manufacturing systems using Petri nets. Liu et al. [18] discussed the application of deterministic and stochastic Petri nets for the performance analysis of a subsea low preventer unit. Kaakai et al. [19] presented the application of hybrid Petri nets in the design of new stations for the transit of passengers, which help the authorities about the safety and security parameters. Khan et al. [20] discussed the applications of Petri nets in deciding the control strategies to reduce the railway. Fecarotti et al. [21] carried out the performance modeling of fuel cell systems using stochastic Petri nets.

Inspired by the great modeling work done by various researchers, an attempt is made in this chapter to use Petri nets as a modeling tool to study the dynamic behavior of the butter oil processing plant. The detailed analysis of the effect of FRRs of various units of the butter oil processing plant on the overall plant availability is performed.

3.2 FORMAL DEFINITION OF PETRI NETS

Algebraically, the Petri nets are represented by five-tuple

$$PN = \{P, T, A, W, M_0\}$$

where

   P represents the finite set of places $\{P_1, P_2, \ldots .P_n\}$
   T represents the finite set of transitions $\{T_1, T_2, \ldots, T_n\}$
   $A \subseteq (PXT) \cup (TXP)$ is a set of directed arcs
   W is a weight function that takes values 1, 2, 3, …
   M_0 is the initial marking.

TABLE 3.1
Graphical Description of Petri Nets

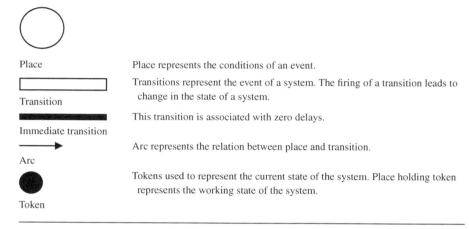

Place	Place represents the conditions of an event.
Transition	Transitions represent the event of a system. The firing of a transition leads to change in the state of a system.
Immediate transition	This transition is associated with zero delays.
Arc	Arc represents the relation between place and transition.
Token	Tokens used to represent the current state of the system. Place holding token represents the working state of the system.

The initial marking represents the initial state of the system. A marking M can be represented by vector $M = \{m_1, m_2...m_i\}$, where m_i represents the number of tokens in the place P_i.

Graphically, the Petri-Net (PNs) are represented by the circles representing the places, rectangles representing the transitions, and arrows representing the arc. There are two arcs, namely, input arc (places to transition) and output arc (transition to places). The dynamic behavior of the system is depicted by the movement of the token from one place to another. The graphical description of Petri net elements is represented in Table 3.1.

3.3   SYSTEM DESCRIPTION

Butter oil or ghee refers to the clarified butterfat obtained mainly from butter after removing all the water and SNF (solids-not-fat) contents. It is the richest source of milk fat and is prepared either by butter or cream. Butter oil production system consists of four subsystems, namely, heater, clarifier, filling, and granulation. Out of these, except for the granulation subsystem, all the units are subject to random failures. The flowchart of the butter oil making process is shown in Figure 3.1.

3.3.1   SYSTEM DESCRIPTION

3.3.1.1   Heater Unit

It consists of a kettle in which the temperature of butter is raised slowly with the help of steam. The final temperature is monitored to be not more than 107°C–110°C until its color is reddish-brown. Ghee along with the residue can settle down for 25–30 minutes in the kettle before filtration. It consists of two units in series. Hence, if one unit fails, the system fails.

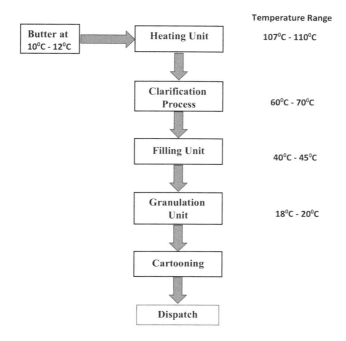

FIGURE 3.1   Flow diagram of butter oil processing system.

3.3.1.2   Clarifier Unit

Clarification is carried out at around 70°C to clarify all the residue particles from ghee. It consists of two units in parallel. Partial failure of this system reduces the capacity of the system. A major failure occurs only when both units fail.

3.3.1.3   Filling Unit

In this section, ghee tins are filled, weighed, and sealed simultaneously. The filling temperature is strictly watched to remain between 40°C and 45°C. There are two filling units. Failure of any one unit reduces the working capacity while the system completely fails when both units break down.

3.3.1.4   Granulation Unit

This subsystem consists of a refrigerating unit where the temperature is maintained between 15°C and 20°C. This section rarely fails and hence has not been considered for analysis.

There are two units in each of the heating, clarifier, and filler. The two heating units are connected in series, and they must run for a system to be operational. The two units of the clarifier and filling units are in a parallel arrangement. Failure of any one of the units of these two leads to the running of the plant at a reduced capacity. The repair of the unit restores the system to its full running capacity. The arrangements of units in the butter oil processing plant are shown in Figure 3.2.

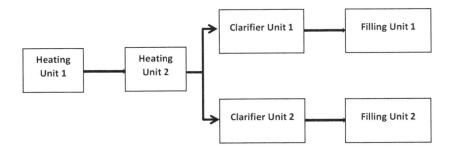

FIGURE 3.2 Schematic arrangement of units of butter oil processing system.

Assumptions for the modeling are as follows:

1. No delay in the repair except the availability of the repair team;
2. Priority of repair is based upon first-come, first-serve basis;
3. Repair is perfect, i.e., system reinstated to its original state;
4. Simultaneous failure can occur in the system; and
5. Repair and failure rates are not dependent on each other.

3.4   PETRI NET MODEL OF THE SYSTEM

To carry out the performance analysis of the butter oil processing system, the interactions among various units of the system are considered. Petri nets are applied to model the interaction between them. If the failures occur simultaneously in the various units of the systems, then repair is done on the first-come first-serve basis, and it also depends upon the number of repair facilities available at the time of failure. Figure 3.3 shows the Petri net model representing the above system. Table 3.2 represents the description of the Petri net model.

3.4.1   Representation of Dynamic Behavior of the System

Figure 3.3 shows the butter oil processing system in upstate. The presence of tokens in places 1, 4, 7, 11, 13, and 16 makes the transitions Tr1, Tr4, Tr7, Tr10, Tr13, and Tr16 enabled. As soon as the stochastic delay of transition Tr1 is achieved, the transition fires lead to the movement of a token from place 1 to place 2. Place 2 representing the waiting for the repair condition of the unit. The token in place 2 enables the immediate transition, which moves the token to place 3, which represents that the unit is under repair. The transition Tr2 gets enabled with the presence of token in place 3. As the stochastic delay becomes equal to repair rate, the transition Tr2 fires which put the token from place 3 to place 1. Similarly, all the units will behave, and depending upon the different guard conditions, the tokens from Place System_Up_State move to System_Down_State. When any one unit of a clarifier or the filling unit fails, the system will run at a reduced capacity, which is represented by a token in the System_Reduced_Capacity place.

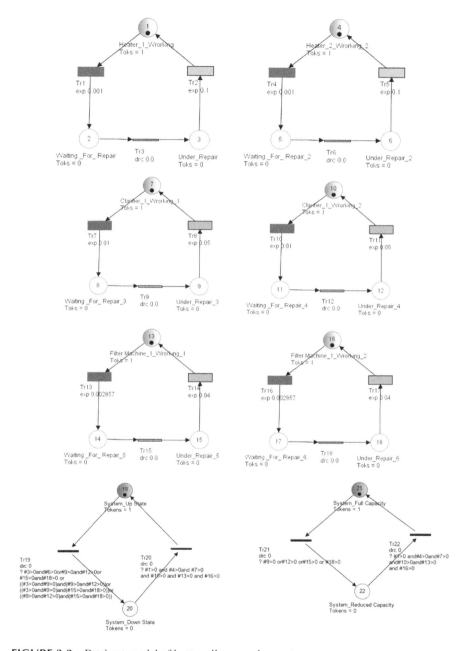

FIGURE 3.3 Petri net model of butter oil processing system.

TABLE 3.2
Description of the Places and Transitions

Places	Description
Heater_1_Working and Heater_2_Working	Places represent the working states of heating units.
Clarifier_1_Working and Clarifier_2_Working	Places represent the working states of clarifier units.
Filler_1_Working and Filler_2_Working	Places represent the working states of filling units.
Waiting_For_Repair_1 and Waiting_For_Repair_2	Places represent the heating unit has failed and is waiting for repair.
Waiting_For_Repair_3 and Waiting_For_Repair_4	Places represent the clarifier unit has failed and is waiting for repair.
Waiting_For_Repair_5 and Waiting_For_Repair_6	Places represent the filler unit has failed and is waiting for repair.
Under_Repair_1 and Under_Repair_2	Places represent the heating unit has gone under repair.
Under_Repair_3 and Under_Repair_4	Places represent the clarifier unit has gone under repair.
Under_Repair_5 and Under_Repair_6	Places represent the clarifier unit has gone under repair.
System_Full_Capacity	Place represents that all the units are working.
System_Reduced_Capacity	Place represents that some of the units are not in working and the system is running with reduced capacity.
System_Up_State	Place represents that the system is in a working state.
System_Down_State	Place represents that the system is in the downstate or the system is not available.

Transition	Description
Tr1, Tr4, Tr7, Tr10, Tr13, Tr16	Represent the time transitions associated with stochastic delay of all the units, and firing of these transitions leads to failure of the corresponding unit. The scholastic delays are related to the failure rates of the units.
Tr2, Tr5, Tr8, Tr11, Tr14, Tr17	Represents the time transitions associated with stochastic delay of all the units, and firing of these transitions leads to repair of the corresponding unit. The scholastic delays are related to repair rates of the units.

3.4.2 Performance Analysis of the System

The performance analysis of the butter oil processing system is carried out using the Monte Carlo simulation of the PN model using the GRIF2020 Petri net module. The simulation evaluates the plant availability. It runs for 10,000 replications for 1 year at a 95% confidence level. The system performance analysis is evaluated by considering the effect of variation of FRRs of the various units the plant availability. Also, the system is evaluated in terms of the plant working at a reduced capacity. The failure and repair data of various units of the system are shown in Table 3.3.

The effect of variation of FRR on the plant availability is shown in Table 3.4. Figure 3.4a shows the effect of variation of FRRs of the heating unit on the availability of the plant. With the decrease in the failure rate, the availability increases by 17%, whereas, with an increase in the repair rate, the availability increases significantly by

TABLE 3.3

Failure and Repair Data for Various Units of Butter Oil Processing System

Units	Failure Data (Exponential Distribution) mean time between failure (in hours)	Repair Data (Exponential Distribution) mean time to repair (in hours)
Heating unit	65	8
Clarifier unit	75	6
Filling unit	85	5

TABLE 3.4

Effect of Variation of FRR of Various Units on Plant Availability

Units	Variation of Failure Rate (Repair Rate)	Plant Availability
Heating unit	0.005–0.025	0.94–0.8
	(0.1–0.4)	(0.7–0.84)
Clarifier unit	0.01–0.033	0.90–0.86
	(0.05–0.2)	(0.81–0.89)
Filling unit	0.0028–0.01	0.88–0.86
	(0.04–0.16)	(0.86–0.88)

20%. Figure 3.4b shows the variation of FRRs of the clarifier unit on the plant availability. As the failure rate decreases, the plant availability increases by 5%. However, with the increase in repair rate, the plant availability increases by 10%. Figure 3.4c depicts the effect of FRRs of the filling unit on the plant availability. Availability increases by 2.5% with the decrease in failure rate, whereas, it increases in the same ratio as the repair rate increases. The above performance analysis suggests that the heating unit has the highest impact on plant availability as compared to other units.

3.5 CONCLUSION

The Petri nets have been observed as powerful emerging tools to study the dynamic behavior of complex systems. In this research, the Petri net-based model for the butter oil processing plant has been developed to carry out the availability analysis of the plant. The PN modeling helps the decision-makers to understand the interaction about the various units of the system. The FRR is exponentially distributed. Upon critical examination of the analysis of the FRRs, it is revealed that the heating unit has a significant impact on plant availability. However, the clarifier and filling units have lesser impact on plant availability. The performance analysis facilitates the plant managers to take decisions related to the maintenance priority and spare part inventory management for achieving the long-term availability of the plant. Apart from these recompenses, the performance analysis can also be used by plant managers for exploring cost-benefit analysis and replacement-related decisions.

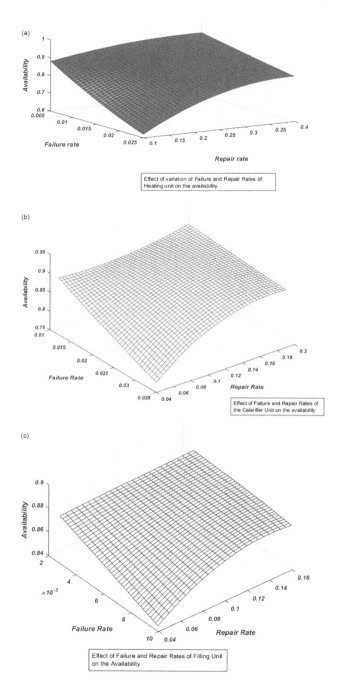

FIGURE 3.4   (a) Effect of variation of FRRs of heating unit on plant availability. (b) Effect of variation of FRRs of clarifier unit on plant availability. (c) Effect of variation of FRRs of filling unit on plant availability.

REFERENCES

1. Mehta M, Singh J, Sharma S. Availability analysis of an industrial system using supplementary variable technique. *Jordan Journal of Mechanical & Industrial Engineering.* 2018;12(4).
2. Zhang X, Gao H, Huang HZ, Li YF, Mi J. Dynamic reliability modeling for system analysis under complex load. *Reliability Engineering & System Safety.* 2018;180:345–51.
3. Byun JE, Noh HM, Song J. Reliability growth analysis of k-out-of-N systems using matrix-based system reliability method. *Reliability Engineering & System Safety.* 2017;165:410–21.
4. Agrawal AK, Murthy VM, Chattopadhyaya S. Investigations into reliability, maintainability and availability of tunnel boring machine operating in mixed ground condition using Markov chains. *Engineering Failure Analysis.* 2019;105:477–89.
5. Pandey P, Mukhopadhyay AK, Chattopadhyaya S. Reliability analysis and failure rate evaluation for critical subsystems of the dragline. *Journal of the Brazilian Society of Mechanical Sciences and Engineering.* 2018;40(2):1.
6. Gupta P, Lal AK, Sharma RK, Singh J. Numerical analysis of reliability and availability of the serial processes in butter-oil processing plant. *International Journal of Quality & Reliability Management.* 2005;22(3):303–16.
7. Garg S, Singh J, Singh DV. Mathematical modeling and performance analysis of combed yarn production system: Based on few data. Applied Mathematical Modelling. 2010;34(11):3300–8.
8. Khorshidi HA, Gunawan I, Ibrahim MY. Data-driven system reliability and failure behavior modeling using FMECA. *IEEE Transactions on Industrial Informatics.* 2015;12(3):1253–60.
9. Murthy C, Mishra A, Ghosh D, Roy DS, Mohanta DK. Reliability analysis of phasor measurement unit using hidden Markov model. *IEEE Systems Journal.* 2014;8(4):1293–301.
10. Bourouni K. Optimization of renewable energy systems: The case of desalination. *Modeling and Optimization of Renewable Energy Systems.* 2012;298:90–116.
11. Imakhlaf AJ, Hou Y, Sallak M. Evaluation of the reliability of non-coherent systems using binary decision diagrams. *IFAC-Papers Online.* 2017;50(1):12243–8.
12. Choi IH, Chang D. Reliability and availability assessment of seabed storage tanks using fault tree analysis. *Ocean Engineering.* 2016;120:1–4.
13. Modgil V, Sharma SK, Singh J. Performance modeling and availability analysis of shoe upper manufacturing unit. *International Journal of Quality & Reliability Management.* 2013;30(8):861–31.
14. Ahmed Q, Khan FI, Raza SA. A risk-based availability estimation using Markov method. *International Journal of Quality & Reliability Management.* 2014;31(2):106–28.
15. Bahl A, Sachdeva A, Garg RK. Availability analysis of distillery plant using petri nets. *International Journal of Quality & Reliability Management.* 2018;35(10):2373–87.
16. Sachdeva A, Kumar P, Kumar D. Behavioral and performance analysis of feeding system using stochastic reward nets. *The International Journal of Advanced Manufacturing Technology.* 2009;45(1):156–69.
17. Patel AM, Joshi AY. Modeling and analysis of a manufacturing system with deadlocks to generate the reachability tree using petri net system. *Procedia Engineering.* 2013;64:775–84.
18. Liu Z, Liu Y, Cai B, Li X, Tian X. Application of Petri nets to performance evaluation of subsea blowout preventer system. *ISA Transactions.* 2015;54:240–9.
19. Kaakai F, Hayat S, El Moudni A. A hybrid Petri nets-based simulation model for evaluating the design of railway transit stations. *Simulation Modelling Practice and Theory.* 2007;15(8):935–69.

20. Khan SA, Zafar NA, Ahmad F, Islam S. Extending Petri net to reduce control strategies of railway interlocking system. *Applied Mathematical Modelling*. 2014;38(2):413–24.
21. Fecarotti C, Andrews J, Chen R. A Petri net approach for performance modelling of polymer electrolyte membrane fuel cell systems. *International Journal of Hydrogen Energy*. 2016;41(28):12242–60.

4 Smart and Reliable Agriculture Application Using IoT-Enabled Fog-Cloud Platform

*Mukesh Kumar Jha Mohit Kumar
and Jitendra Kumar Samriya*
NIT Jalandhar

CONTENTS

4.1 INTRODUCTION

Agriculture, the largest contributor to a country's economy, has largely been untouched by the benefits of ever-evolving, ever-improving technology. The main reasons can be attributed to the lack of technical expertise, low bandwidth availability in rural areas, infrastructure constraints, high latency applications, etc. The

DOI: 10.1201/9781003140092-4

emergence of IoT cloud computing has opened an array of possibilities. Huge volumes of data can now be collected and sent to applications without the need for any physical infrastructure. This can help improve soil management, improve irrigation planning, and track pest movement, ultimately resulting in higher productivity for the farmers. The ready availability of resources on the cloud has made it possible to develop low-cost applications.

Smart agriculture involves using technology to augment farm-related decision-making to expand yield by making the best possible use of resources at hand. Agriculture-related decisions are taken based on factors such as temperature, humidity, soil composition and pH, pest movement and water level. These metrics are collected by numerous IoT-enabled sensors and other devices and transmitted to processing centres in real time. Traditional agricultural applications based on cloud computing are inadequate in front of such volumes of data. The lower network bandwidth availability in rural areas adds to the problem. As more and more devices are connected, there is a need to move the data processing points closer to the data generation points, to help farmers make timely decisions. A fog computing-based solution can have better utilization of the bandwidth and lower battery consumption on IoT devices, and provide low latency [1]. A fog device, placed close to the farm, has reasonable computing and storage capacity to collect and process data collected from the local farm sensors. The fog node then communicates its results to the farmer through a web/mobile application. Additionally, the local data collected from different fog locations at the centralized server can then be used by geoagri researchers to model crop patterns, irrigation patterns and by the government to prepare an agricultural census for devising appropriate policies [2].

4.1.1 Issues with Agriculture

The way the world population is growing will pose a big challenge to feed them properly in a few decades. Vast industrial development, urbanization and many more factors are very much responsible for the decrease in cultivable land. There are also trends of water shortage across the globe which may affect agriculture badly. So it is the need of the hour to move towards smart agriculture from traditional one. In the current scenario, it can be said that the following issues are most prevalent in agriculture:

- Water crisis
- Cultivable land
- Efficient use of fertilizers
- Climate change

4.1.2 Need of Technology in Agriculture

Today most countries are adopting new technologies and policies related to smart agriculture. Precision farming and vertical farming are examples of smart agriculture. In India, "per drop more crop" policies depict the government's sensitivity towards the water shortage. Recently, five memorandum of understanding have been signed between Government of India and private partners for taking forward the "digital agriculture" and connecting agriculture with technology. In today's era, smartness in agriculture can be synonymously used as digital agriculture. Digital technologies have shown their potential in industrial development, mechanization, etc., where all the inspection and the quality control can be done digitally without human intervention. Digital technology has full potential to bring the second wave of the green revolution. Smart agriculture can be implemented for dairy technology, poultry, animal husbandry, fisheries, beekeeping, etc.

The various technologies that can enhance the agriculture sector are shown in Figure 4.1. The term digital technologies comprise Internet of Things (IoTs), cloud computing, edge computing and artificial intelligence like technologies. The usage of

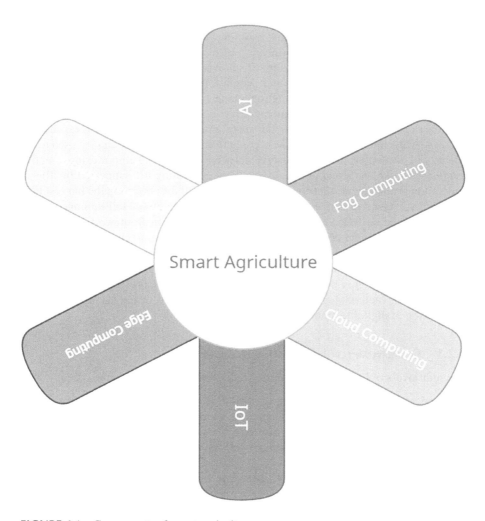

FIGURE 4.1   Components of smart agriculture.

these technologies may lead to an increase in production with less manual intervention and finally may lead to an increase in the farm income.

In each activity of farming, the digital technologies can be implemented. Few farming activities with applications are as follows:

- Soil analysis
- Irrigation
- Fertilization
- Air monitoring
- Water monitoring
- Pest control

4.2 IoT (INTERNET OF THINGS) IN AGRICULTURE

IoT arrangements are centred around assisting ranchers with shutting the stockpile request hole, by guaranteeing significant returns, productivity and insurance of the climate. The methodology of utilizing IoT innovation to guarantee ideal use of assets to accomplish high harvest yields and decrease functional expenses is called accuracy agribusiness. IoT in farming advancements contain specific gear, remote network, programming and IT administrations.

Smart cultivating in light of IoT advances empowers cultivators and ranchers to decrease waste and improve usefulness going from the amount of compost used to the quantity of excursions the homestead vehicles have made, and empowering effective use of assets like water and power. IoT brilliant cultivating IoT supported Intelligent cultivation arrangement is a framework that is worked for observing the harvest field with the assistance of sensors (light, mugginess, temperature, soil dampness, crop well-being, and so forth) and robotizing the water system framework. The ranchers can screen the field conditions from anyplace. They can likewise choose among manual and robotized choices for making important moves in view of this information. For instance, assuming the dirt dampness level abatements, the rancher can convey sensors to begin the water system. Smart cultivation is effective in compared to the tradition methodologies of farming. These are a few examples of a smart way of inclusion of digital technologies in agriculture.

4.2.1 FRAMEWORK FOR IoT-BASED AGRICULTURE

The four layers of IoT-based agriculture framework is shown in Figure 4.2.

Physical layer: It is the first layer of the IoT architecture device connectivity. The different sensors and actuators are connected to gather information about temperature, pressure, moisture, CO_2 level, wind level, soil information, etc. At this layer, the data are collected and passed to the above layer for further analysis.

Edge layer: Edges are the actual "things" of IoT. The edge layer is responsible for connecting devices locally, collecting the data and manage the connection to the server. The real-time data visualization, real-time monitoring, data pre-processing, and all the activities take place at the edge layer.

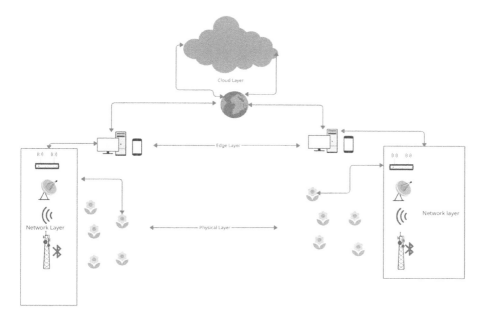

FIGURE 4.2   IoT framework for digital agriculture.

Network layer: This layer is known as the backbone of IoT. All types of network connectivity are handled at this layer. 3G, 4G, Wi-Fi, and infrared are
managed at this level. The communication between the sensors and different devices is taking place with the help of this layer only.

Cloud layer: Since different devices and sensors emit enormous amount of
data, for storing, decision-making and handling big data, the cloud layer
plays a pivotal role in giving the illusion to the users about the bulk amount
of storage space and decentralized decision-making.

4.2.2   IoT-Based Precision Agriculture

Today, India ranks second in the world in farm output. Most of the cultivated land
depends on monsoons. Irrigation accounts for 80%–85% of water usage in India in
which 60% of water gets wasted, so it is needed to do smart irrigation through IoT-
based sensors. Modern information technologies are very much helpful in dealing
with efficient water usage for irrigation.

There are lots of benefits in IoT-based precision farming, and some of them are
as follows:

• Enhancing productivity in agriculture
• Prevention of soil degradation in cultivable land
• Reduction of chemical use in crop production

TABLE 4.1

Various Work Related with IoT Approaches in Agriculture

Authors	Application Domain	Description
W. Zhao, X. Wang, B. Qi and T. Runge	Mesh Simultaneous Localization and Mapping algorithm (Mesh-SLAM) and Internet of Things (IoT)	Used Mesh-SLAM algorithm for mapping and localization precision, accuracy, and yield prediction error, IoT system for modularized, easy adding/ removing new functional modules or IoT sensors and finally calculated the trade-off between cost and performance which widely augments the feasibility and practical implementation of this system in real farms [3].
A. D. Boursianis et al.	The AREThOU5A IoT platform	Proposed the architecture of an intelligent irrigation system for precision agriculture, the AREThOU5A IoT platform [4]. We describe the operation of the IoT node that is utilized in the platform.
A. Vangala, A. K. Das, N. Kumar and M. Alazab	IoT-based agriculture: blockchain perspective	Developed block chain-based security architecture, conducted cost analysis, comparative analysis and highlighted drawbacks in existing research [5].

- Efficient use of water resources
- Efficient farm management
- Improvement in quality and reduction in the cost of production in crops.
  - Wired or wireless automation in micro irrigation systems specifically in surface-covered cultivation
  - Actuator-based operational automation in supply and demand management in on-farm water management

The decision of "when to irrigate" is dependent on an irrigator's experience. The application of irrigation water to each basin is time based, which could lead to lower application efficiencies.

4.2.2.1 Related Work with IoT Approaches in Agriculture

Nowadays, various researches are going on for finding the application of IoT in agriculture. Some of the approaches of researchers are mentioned in Table 4.1.

4.3 CLOUD AND FOG COMPUTING

In simple terms, cloud computing means using remote servers for computation or storage instead of developing and maintaining local infrastructure. The advent of cloud computing has made high-end computing power accessible to the general public and at the same time, it has provided opportunities to run highly scalable, reliable and fast applications without the overhead of maintaining local servers. As per

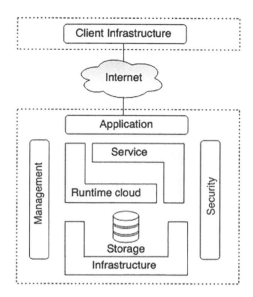

FIGURE 4.3   Architectural view.

our need, we can avail services in these modes: Infrastructure as a Service (IaaS), Platform as a Service (PaaS) or Software as a Service (SaaS). Services such ascloud formation by amazon web service have even made it possible to define your infra-structure as code that automatically manages your servers according to run time parameters. The pay-as-you go model of costing lures more and more people to go for cloud-based services as you are charged only for the number of resources used by you. The architectural view of cloud is shown in Figure 4.3.

Cloud computing is the arrangement of PC or IT framework through the Internet. That is the provisioning of shared assets, programming, applications and adminis-trations over the web to satisfy the versatile need of the client with least exertion or communication with the specialist organization. India is perhaps the biggest maker of food sources, grains and different items, yet at the same time, horticulture and its creation interaction are decentralized, unsophisticated and obsolete strategies being trailed by the ranchers, along with a few imperatives of the ranchers and moderniza-tion is exceptionally sluggish. This will contrarily affect the rancher's financial cir-cumstances and also the public pay of the country. This bottleneck can be eliminated with the implementation of cloud computing office in farmland.

The brought-together area must be set up to store every one of the significant information. It may include the distinct soil, climate, crop and farmers related infor-mation at one place for the easy accessibility of information.This information can be accessed by the end clients such as ranchers, specialists, advisors and analysts effectively any time from any area through the gadgets that are associated with the cloud framework. As more and more smart devices which can communicate over the Internet are popping up daily, the traffic load on limited bandwidth networks has increased. Also, with IoT, the sensors and devices are collecting and exchanging

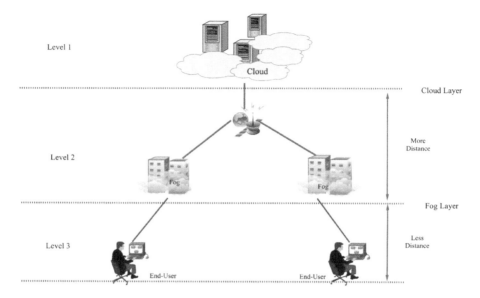

FIGURE 4.4   Architectural view of fog computing.

humungous volumes of data. The conventional cloud computing-based architectures suffer latency in applications involving huge data transactions.

Fog computing is what provides the solution. The name "fog" is derived from the architectural feature that the data are now processed closer to the source just as "fog" lies closer to humans than the "clouds". The fog devices are smart devices that can process data collected from local sensors. These devices then communicate their results and findings to the end user and occasionally to the main server on the cloud. Architectural view of fog computing is shown in Figure 4.4.

The real-time data collected from IoT devices and sensors are not communicated to the cloud, thus reducing network traffic. The communication between cloud servers and fog devices is limited to receiving software updates, sending periodic reports, changing global parameters, etc. The reduced latency thus aids real-time decision-making in applications that are time-sensitive. Fog computing is a disseminated stage that offers execution, stockpiling and correspondence offices between the IoT gadgets and the cloud figuring stockpiling region. The fog layer is currently considered as a significant expansion of the cloud as it gives many supporting and accommodating qualities, for example, edge position, position mindfulness and little reaction delay. The possibility of the fog was proposed mostly to support the end-client gadgets with administrations, like the ongoing applications (e.g. web-based games, online gatherings and expanded reality). In certain applications, servable hubs are conveyed on account of the wide geo distribution of the central concern, for example, field observing specifically. Mist hubs can be different in their inclination and can be conveyed in a broad scope of conditions. For such explanation, we really want to track down strategies for proficient appropriation of the models and errands.

4.4   RELATED WORK

A lot of researches and studies are going on that basically aims towards the application of technologies like Fog and IoT in agriculture for the betterment of cultivation of crops and farmer's economic conditions. These works are as follows:

   i. **Agrifog:** A fog computing and IoT-based model was developed by Sucharitha, V., Prakash, P. and Ganesh Neelakanta Iyer. The model makes use of sensors/actuators and smart devices at the fog layer. The model was simulated on iFogSim with both cloud-based and fog-based configurations. The results indicated lower network usage, and lower latency in fog-based configuration is shown in Figures 4.5 and 4.6.
  ii. **Precision farming model with IoT and Fog computing context:** Developed by Francisco Javier Ferrández-Pastor, Juan Manuel García-Chamizo and Mario Nieto-Hidalgo, this model discusses a user-centric computation model which uses machine learning models to automate irrigation and fertilizing processes. The model discusses a three-layer communication architecture that assists the training of models at different levels with data sets collected from different farms (fog nodes). This model has been proposed at EU parliament, and it has started getting adopted in some parts of Europe.
 iii. **SmartHerd management:** IoT-assisted fog computing platform for smart dairy farming for animal behaviour analysis and health monitoring. The system analyses animal movements, steps taken, rest time and other variables in an ML model to flag erratic behaviour. The model was tested in Ireland on a herd of 150 cows. Although the cost of initial set-ups was high,

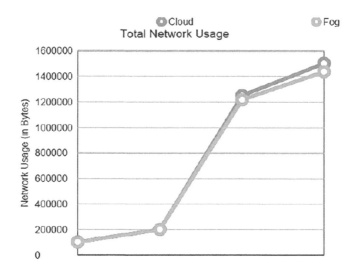

FIGURE 4.5   Network usage.

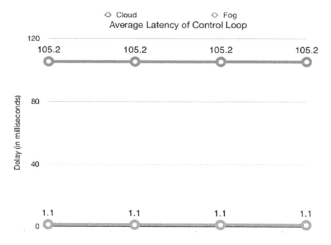

FIGURE 4.6   Average latency of control loop.

TABLE 4.2
Various Smartphone Apps with Agricultural Applications [7]

App Name	Application Domain	Features
EcoFert	Soil health	Helps calculate optimum level and combination of fertilizers based on soil nutrient requirements and current market prices of fertilizers.
VillageTree	Pest management	It uses a crowdsourcing approach to gather pest attack reports from farmers and alerts those who may be affected well advance in time with recommended pesticides.
PocketLAI	Irrigation	It uses leaf area index (LAI) to determine plant's water requirements based on evapotranspiration. Leaf images are collected either by fixed sensors or using smartphone camera.
WheatCam	Crop insurance	It uses picture-based insurance (PBI) to improve the quality of crop insurance claims. It aids in the better settlement of insurance claims as they are backed by data collected over the period of crop damage.
Agrimaps	Land management	This app provides special data visualization for site-specific land use and crop pattern recommendations.
BioLeaf	Crop health	It monitors crop health using leaf foliage damage data with or without leaf border damage using on-field cameras or smartphone cameras.

the results showed 84% reduction in network traffic from individual farms to the cloud. This also resulted in better response time as computation and data communication from individual sensors were limited to the local fog node over a lighter communication protocol, MQTT [6] (Table 4.2).

4.5   USE CASES

The ultimate goal is to aid the farmer in making better decisions, based on both local and global data. The primary use cases are as follows:

- Farmers can decide on composition, quantity and timing of adding fertilizers to supply essential nutrients and maintain a suitable soil profile for the crop.
- Appropriate water level can be maintained, and farmers can decide on when and how much to irrigate. This can be aided by weather forecast module to warn of potential rainfall.
- Farmer can decide on which crop to grow based on soil profile, humidity, temperature and other environmental conditions.
- The reports and data collected from different edge locations (a set of fog devices) can be used to train mathematical models of crop patterns, harvesting patterns, irrigation patterns, finding best time to sow, etc. [8].
- The global data can be used to prepare land use patterns, soil map, agriculture census, etc. by the government agencies, thus impacting the policies [9].

Figure 4.7 shows the diagrammatic representation of farmers', government's, and researcher's roles in agriculture based on IoT.

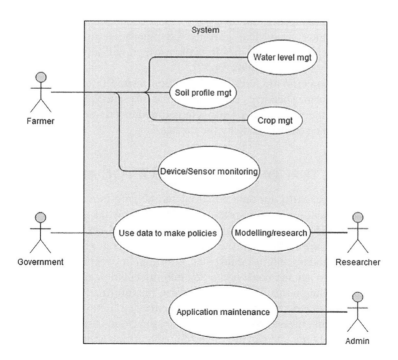

FIGURE 4.7 Diagram representing farmer, government, researcher's role in agriculture based on IoT.

4.5.1 IMPACT ON FARMERS' INCOME

1. **Climate conditions monitoring:** Continuous monitoring of climate conditions can provide better ways of planning future activities, which consequently improves agricultural productivity. Weather stations are installed across the field to monitor different weather parameters. Some of these parameters are temperature, humidity, wind direction, air pressure, etc.
2. **Soil patterns:** Soil patterns to be observed are soil humidity, moisture, fertilization and temperature. Soil humidity and moisture sensors are installed in soil. Soil pattern report helps in selecting suitable fertilizers, thus resulting in improved crop yield. Also, identification of contaminated soil can save field from crop loss and over-fertilization.
3. **Pest and crop disease monitoring:** Prediction of crop disease at early stages of crop growth can help the farmers to generate more yield by saving crops from pest attacks. Manual disease detection by crop experts is a very expensive and time-consuming process, instead of that automatic disease detection is very accurate and cheaper comparatively.
4. **Irrigation monitoring system:** The irrigation system can be optimized by monitoring soil patterns and weather patterns. Various water parameters monitored are pH, oxygen, temperature and conductivity [10]. These modern techniques reduce irrigation costs and limit water resources usage.
5. With better monitoring and insights, farmers can explore new agricultural techniques such as vertical farming, greenhouse farming, phenotyping, hydroponic and autonomic drip irrigation, which will ultimately help them face adversity on their staple crop [7].
6. **Crop insurance:** With detailed information related to crop, crop damage claim for a farmer becomes easier. Also, the insurance provider can install their own cameras and other sensors to gather data pre- and post-damage to settle insurance claims made by the farmers.

4.6   ARCHITECTURE OF INTEGRATION OF FOG AND CLOUD

Cloud is utilized generally for data handling and helping in making the right decisions. The judgement has a more profound idea than this taken by the fog layer. It could be postponed and can be founded on collective information. In the association, the fog nodes are liable of getting the IoT gadgets' continuous information and store them momentarily and just convey an information overview to the cloud. The cloud stage job is to get and consolidate Fog hub information and then, at that point, perform examination on this information. At last, the information is sent back to the fog hubs from the cloud after this examination either straightforwardly or a control signal in view of this investigation. In this sense, the cloud is treated as the director for the entire framework and can use the profound learning approaches and information examination to get a bird view look. For instance, on this fog progressive system could be business building the executives, on the whole consumption of water, power expected for poultry, fertilizers and pesticides.

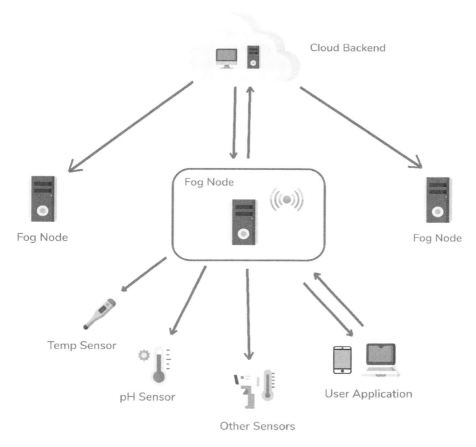

FIGURE 4.8   Architecture of integration of fog and cloud.

The fog shown in Figure 4.8 handles the time-delicate handling, while the cloud handles the working and business-related information handling. Models on this situation could incorporate agribusiness noticing and portable organization speedup. Nonetheless, the fog hubs can be allocated a few observing and control liability uncommonly for insurance control. The fog registering designers ordinarily use fog hubs to compose IoT applications, where they are close to the organization edge. As per the kind of information, the fog IoT application conveys the information to the legitimate layer for more examination, whenever required. For instance, the exceptionally time-delicate information could be better examined on the fog hub adjoining to the IoT gadgets creating the quick and proper control information, while the non-critical information is better shipped off the cloud for more examination and long-lasting stockpiling.

Crop field monitoring inspect where there is a requirement for the cloud for information investigation For this situation, fog components are associated straightforwardly to the cloud as displayed in the figure. Numerous sensor hubs are expected to gauge, for instance, light, moistness, soil dampness, temperature and power. The fog components can present capacities like plant development checking, update for collect assembling, manure showering, pesticides and water system. Cloud provides business information that yield in better profit. It can give the fog hubs weather conditions conjecture, which oversees the water system plan set by the fog hub. The diagram shown in Figure 4.8 shows how the integration of cloud and fog can be helpful in agriculture using the various fog nodes, cloud and sensors.

The architecture shown in the diagram consists of following components:

i. Cloud backend
   • Manage and maintain the fog nodes. Routine health check-ups, state monitoring and alerting the admin about any faulty/non-responding fog nodes.
   • Push updates to the application running on fog nodes and change global parameters affecting decision-making at the farm level.
   • Collect periodic data from fog nodes and save it in data repositories. Aid government agencies and researchers by providing access to the data.

ii. Application running on fog device
   The application will have a modular structure with each module handling a different responsibility. This will enable easy extension in future to add more functionalities.
   • Maintain and monitor sensors/routers on the farm. Alert the farmer about any malfunctioning device by performing routine health check-ups.
   • Collect data from sensors, process it and provide results to the farmer through web/mobile application.
   • Receive and install updates to the application. Change the model/parameters that process the data, as received from the centralized backend.
   • Send periodic reports and data to a centralized server.

iii. User application
   • A web or mobile application to be used by the farmer.
   • Connect to the local fog server and display the results to user in an easy to consume manner.
   • Display the central dashboard of devices and their health/battery.
   • A weather module that provides the local weather forecast to the farmer.
   • Collect periodic data from the farmer and communicate it to the local fog server.

The water system checking and controlling framework can follow the Hierarchal Fog and Cloud Architecture, where every area is the obligation of a fog component. The different locales are then the obligation of a higher fog hub which screens these different districts. The entire framework conveys its information for the cloud to settle on higher choices and for any reports that help the country leader in accepting its future choices as displayed in the figure.

Illustration of sensors are temperature and moistness sensor, soil dampness sensor, attractive float sensor for water level pointer, and ebb and flow level of water in the waterways sensor. The cloud can make a full water system plan while the fog upper level can settle on more modest choices while checking different segments of fields. The lower-level fog hubs have direct cooperation with the sensors/actuators of a more modest region of the field planted with various kinds of harvests which settle on its choice as regards to this sort of yield more exact. In view of the assessment of the water expected for the harvest type, the speed and measure of water are changed according to the flood and field.

IoT has made critical improvements in farming administration in agriculture. It permits generally agrarian gadgets, and the gear has to be connected together to settle on the fitting choice in the water system and compost supply. The shrewd frameworks improve the exactness effectiveness of gadgets that check plant development and, surprisingly, raise domesticated animals. Remote sensor networks are utilized to gather information from various detecting gadgets. Moreover, cloud administrations are likewise fundamental to be incorporated with IoT to investigate and handle the far-off information that works with decision production to execute the best choices. Smart farming requires utilizing ICT, ground sensors and control frameworks introduced on robots, independent vehicles and other mechanized gadgets. The accomplishment of savvy frameworks relies upon high velocity web, progressed cell phones and satellites to give pictures and location. IoT is utilized as constant to follow and analyse leaf illnesses that hinder crop development utilizing many satellite pictures and sensors situated in ranches.

4.6.1 CATTLE OBSERVATION

IoT applications assist ranchers in gathering information regarding the area, prosperity and strength of their dairy cattle. These data help them in recognizing the state of their domesticated animals. For example, observing creatures that are debilitated, thus, that they can isolate from the group, forestalling the spread of the infection to the whole dairy cattle. The achievability of farmers to find their dairy cattle with the assistance of IoT-based sensors helps in cutting down work costs by a significant sum.

Organizations make use of IoT for overseeing dairy ranch tasks. They sent off savvy neck collars for cows. A painless checking framework run progressed restrictive calculations joining dairy science, master information and AI to convey significant data like temperature, movement, rumination and conduct for any well-being alarms, infection indications, oestrus location and feed enhancement.

The various applications of IoT sensors in the agriculture sector are categorized into various categories [11] as shown in Figure 4.9.

4.6.2 OBSERVING ENVIRONMENT CONDITIONS

Weather condition stations outfitted with savvy sensors can gather climate information and send valuable data to a rancher. In addition, the data are dissected by unique programming, and the inspection assisted by rancher assists him with having a definite conjecture and stay away from crop misfortunes [12].

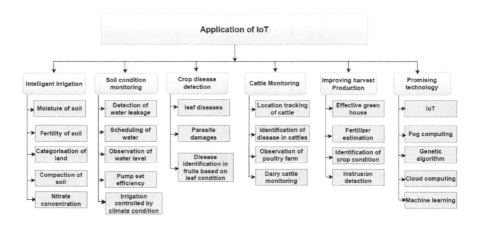

FIGURE 4.9   Application of IoT sensor in agriculture.

For instance, all weather monitoringis an IoT-based rural venture that offers ranchers exceptional programming and gear for checking atmospheric conditions, which cautions the early admonition of outrageous temperatures, ice and stormy climate on your homestead fields.

4.6.3   NURSERY AUTOMATION

As well as obtaining ecological information, weather condition stations can consequently change the circumstances to match the given boundaries and to give the most fitting condition to every nursery.

GreenIQ is likewise an intriguing item that utilizes brilliant farming sensors. It is a brilliant sprinkler regulator that permits you to deal with your water system and lighting frameworks from a distance.

4.6.4   CROP CHECKING

As on account of atmospheric condition monitoring sensors for crop checking likewise gather all data like yield well-being, moistness, precipitation, temperature and different boundaries. In the event that there are any deviations, ranchers might recognize them ahead of time and make suitable moves. Likewise, sensors assist ranchers with the alternatives for the crop. [13]. Semios, one of the most well-known IoT projects for observing harvests, furnishes ranchers with controller of environment, bug and sickness checking.

4.6.5   DRONES

In accuracy horticulture, drones have a scope of employments from soil and harvest field examination to planting and pesticide splashing. Robots can be utilized with various imaging advances like hyperspectral, multispectral and warm that can furnish the ranchers with time and site-explicit data in regard to trim well-being, contagious contaminations, development bottlenecks, and so on.

Drones have been utilized economically in farming since the mid-1980s, yet with the improvement of correspondence innovation and the extended utilization of IoT, the utilization of automated airplane has become vital. It can fill various roles that lead to working on horticultural practices. Some of the robot tasks include checking water system crop well-being, planting, crop showering, crop examination and soil investigation [14]. Likewise, the robot furnished with a few sensors, 3D cameras, warm, multi-ghastly, and optical imaging cameras can be utilized to screen crop conditions and sicknesses, plant well-being markers, vegetable thickness, pesticide prospecting, manure, shelter cover planning, field forecast, plant count, plant tallness estimation, field water planning, exploratory reports and nitrogen estimation. Moreover, it can screen the condition of plants in light of some vegetative lists that can be straightforwardly determined by multi-ghostly pictures which are viewed as one of the most popular [15].

Robots can likewise distinguish drier locales in a field and measures can then be taken to flood such districts with better methods. Accuracy horticulture gives ranchers such substantial data that empower them to settle on informed choices and use their assets effectively. The implementation of IoT, Fog and Cloud in the agriculture intends to help farmers and agriculture investors to take the right judgement. The use of IoT reduces the expense of crop production. It depicts the efficiency and soil supplement the board in brilliant cultivating. The high supplement level of soil guarantees the upkeep of nature of yield. The dirt disintegration prompts the absence of supplement level in the soil [16]. The indicating of plants which can fix the nitrogen on soil ought to be utilized to expand the nitrogen level in soil. The farmers ought to stay away from the planting of a similar harvest on the particular yield multiple times. The appropriate approach to testing on the components of soil upgrades the supplement level to build the creation. The appropriate utilization and follow-up of composts will work on the supplement level of soil. The planting of same species in a similar harvest yield will lessen the supplement level of soil [17]. The uniform harvest fosters the construction of root to expand the development of yield. The uniform harvest expands the strain to claim the good productivity.

Smart farming can reduce the cost of production in various ways like efficient usage of irrigation system, smart usage of pesticides and efficient use of fertilizers. Smart usage of pesticides characterize that the harvests need pesticide to keep away from yield decrease due to pesticides [18]. The farmers need to foresee and utilize the pesticide as indicated by the need; however, the utilization of pesticide isn't ideal. Pesticides are over utilized and squandered. This can be tried not to utilize the savvy gadget which will screen and utilize the pesticide as per the need of the yield. Manure assists the ranchers in getting the great-quality crossbreed crop. Over- and under-utilization of compost is really occurring. Savvy IoT gadgets can be utilized to screen the yield to distinguish the need of the manure and appropriately it is utilized which will enhance the utilization of compost and diminish the expense [19]. The savvy water system framework coordinates to further develop the efficiency progressively. Crop water system is a craftsmanship and ought to be appropriately done to diminish the wastage of water. The examination is done for a brilliant water system. The right strategy for appropriate water system will improve the efficiency of farming. IoT-controlled irrigation system reduces and controls the wastage of water.

4.7 CONCLUSION AND FUTURE RESEARCH DIRECTIONS

There are lots of techniques that have been proposed by the researchers to resolve the issues of farmers and implement the smart farming. Most of the farmers still using traditional farming methods and do not adopt the latest technology-based smart and reliable farming to gain the maximum profit. Hence, in this chapter, we have proposed the smart farming approach using the thrust technologies like Internet of Things, emerging computing paradigm named fog and centralized server for storing and processing the huge data cloud computing platform. The proposed smart farming architecture is based upon the IoT sensors that collect the important information like humidity, air condition, weather condition and water level that are helpful in improving the conditions of Indian farming. Further, the collected information is processed over the fog platform without latency There are various IoT sensors used in the agriculture that have been categorized in this chapter. In the future, we try to use machine learning approaches to predict the soil conditions so that farmer can utilize the land for better crop production based on the geographical conditions of that particular land.

REFERENCES

1. Sucharitha V, Prakash P, Ganesh Neelakanta I. AgriFog-A fog computing based IoT for smart agriculture. *IJRTE*. 2019:2277–3878.
2. Precision Farming model with IoT and Fog computing context by Francisco Javier Ferrández-Pastor, Juan Manuel García-Chamizo and Mario Nieto-Hidalgo. https://www.researchgate.net/publication/325419033_Precision_Agriculture_Design_Method_Using_a_Distributed_Computing_Architecture_on_Internet_of_Things_Context.
3. Zhao W, Wang X, Qi B, Runge T. Ground-level mapping and navigating for agriculture based on IoT and computer vision. *IEEE Access*. 2020;8:221975–85.
4. Boursianis AD, Papadopoulou MS, Gotsis A, Wan S, Sarigiannidis P, Nikolaidis S, Goudos SK. Smart irrigation system for precision agriculture—The AREThOU5A IoT platform. *IEEE Sensors Journal*. 2020;21(16):17539–47.
5. Vangala A, Das AK, Kumar N, Alazab M. Smart secure sensing for IoT-based agriculture: Blockchain perspective. *IEEE Sensors Journal*. 2020;21(16):17591–607.
6. Taneja M, Jalodia N, Byabazaire J, Davy A, Olariu C. SmartHerd management: A microservices-based fog computing–assisted IoT platform towards data-driven smart dairy farming. *Software: Practice and Experience*. 2019;49(7):1055–78.
7. Ayaz M, Ammad-Uddin M, Sharif Z, Mansour A, Aggoune EH. Internet-of-Things (IoT)-based smart agriculture: Toward making the fields talk. *IEEE Access*. 2019;7:129551–83.
8. Nandhini S, Bhrathi S, Goud DD, Krishna KP. Smart agriculture IOT with cloud computing, fog computing and edge computing. *International Journal of Engineering Advanced Technology*. 2019;9(2):3578–82.
9. Kunal S, Saha A, Amin R. An overview of cloud-fog computing: Architectures, applications with security challenges. *Security and Privacy*. 2019;2(4):e72.
10. Givehchi O, Jasperneite J. Industrial automation services as part of the Cloud: First experiences. *Proceedings of the Jahreskolloquium Kommunikation in der Automation–KommA, Magdeburg*. 2013.
11. Said Mohamed, E, Belal, AA, Abd-Elmabod, SK, AEl-Shirbeny, M, Gad, A, Zahran, MB. Smart farming for improving agricultural management. *The Egyptian Journal of Remote Sensing and Space Science*. 2021;24(3, Part 2):971–81.

12. Ray PP. Internet of things for smart agriculture: Technologies, practices and future direction. *Journal of Ambient Intelligence and Smart Environments.* 2017;9(4):395–420.

13. Friha O, Ferrag MA, Shu L, Maglaras LA, Wang X. Internet of Things for the future of smart agriculture: A comprehensive survey of emerging technologies. *IEEE/CAA Journal of Automatica Sinica.* 2021;8(4):718–52.

14. Elijah O, Rahman TA, Orikumhi I, Leow CY, Hindia MN. An overview of Internet of Things (IoT) and data analytics in agriculture: Benefits and challenges. *IEEE Internet of Things Journal.* 2018;5(5):3758–73.

15. Zhang X, Zhang J, Li L, Zhang Y, Yang G. Monitoring citrus soil moisture and nutrients using an IoT based system. *Sensors.* 2017;17(3):447.

16. Agrawal H, Dhall R, Iyer KS, Chetlapalli V. An improved energy efficient system for IoT enabled precision agriculture. *Journal of Ambient Intelligence and Humanized Computing.* 2020;11(6):2337–48.

17. Turner LW, Udal MC, Larson BT, Shearer SA. Monitoring cattle behavior and pasture use with GPS and GIS. *Canadian Journal of Animal Science.* 2000;80(3):405–13.

18. Ibrahim H, Mostafa N, Halawa H, Elsalamouny M, Daoud R, Amer H, Adel Y, Shaarawi A, Khattab A, ElSayed H. A layered IoT architecture for greenhouse monitoring and remote control. *SN Applied Sciences.* 2019;1(3):1–2.

19. Zamora-Izquierdo MA, Santa J, Martínez JA, Martínez V, Skarmeta AF. Smart farming IoT platform based on edge and cloud computing. *Biosystems Engineering.* 2019;1(77):4–17.

5 A Reliable Inventory Model for Deteriorating Items with Multivariate Demand under Time-Dependent Deterioration

Sanjay Sharma and Anand Tyagi
Ajay Kumar Garg Engineering College

Sachin Kumar
KIET

Richa Pandey
Graphic Era Hill University

CONTENTS

5.1   INTRODUCTION

The first attempt to obtain optimal replenishment policies for deteriorating items was made by Ghare and Schrader [1], who derived a revised form of the economic order quantity (EOQ) model assuming exponential decay. Pakkala and Achary [2] presented a two-level storage inventory model for deteriorating items with bulk release rule. In the above-mentioned models, the demand rate was assumed to be constant. Subsequently, the ideas of time-varying demand and stock-dependent demand were considered by some other authors, such as Goswami and Chaudhuri [3], Bhunia and

DOI: 10.1201/9781003140092-5

Maiti [4], Benkherouf [5] and Kar et al. [6]. Zhou and Yang [7] studied stock-dependent demand without shortage and deterioration with quantity-based transportation cost. Wee et al. [8] considered a two-warehouse model with constant demand and Weibull distribution-based deterioration under inflation. Hsieh et al. [9] developed a deterministic inventory model for deteriorating items with two warehouses by minimising the net present value of the total cost. In that model, they allowed shortages that were completely backlogged. Ghosh and Chakrabarty [10] suggested an order-level inventory model with two levels of storage for deteriorating items. Shortages were allowed and fully backlogged. Jaggi and Verma [11] developed a two-warehouse inventory model with linear trend in demand under the inflationary conditions with constant deterioration rate. This paper presents an order-level inventory model for deteriorating items with two warehouses. Demand rate is taken as stock dependent. Sharma et al. [12] presents an inventory model for decaying items, considering the multivariate consumption rate with partial backlogging. Sharma et al. [13] developed a two-warehouse production policy for different demands under volume flexibility. Sharma et al. [14] present an economic production quality (EPQ) model for deteriorating items with price-sensitive demand and shortages in which production is demand dependent.

Khanna et al. [15] discussed optimising preservation strategies for deteriorating items with time-varying holding cost and stock-dependent demand. Chauhan and Tayal [16] developed an order quantity scheme for ramp-type demand and backlogging during stock-out with discount strategy. A model in which first time demand was assumed as a function of backlog, Rout et al. [17] presented a production inventory model for deteriorating items with backlog-dependent demand.

A detailed review based on the deterioration of the instant and non-instant items was analysed by Saxena [18], in which the review on EPQ models for instantaneous and non-instantaneous deteriorating items was developed. Singh et al. [19] presented an EPQ model using lifetime items with multivariate demand markdown policy under shortages and inflation. Mohan et al. [20] developed an inventory model for decaying items with Pareto distribution, time-dependent demand and shortages.

In the present study, the main focus on demand is considered, and to avoid unrealistic ups and downs in the demand, we have combined all the above-mentioned factors to make this study more realistic. In this model, we have focused on the pattern of demand, so this study becomes more useful for the business prospects. The model is developed for deteriorating items, where the rate of deterioration is assumed to be time dependent. The demand rate is taken as a function of the time-dependent constant and selling price. The shortages are not allowed. A numerical example and a sensitivity analysis are shown to illustrate the model and its significant features.

5.2 NOTATIONS AND ASSUMPTIONS

The following notations are used throughout the chapter:

$I(t)$: The inventory level at any time, $t \leq 0$
T: Cycle length (time units)
$D(t)$: The demand rate

P: Unit selling price per unit
S: Shortages cost per unit
A: Ordering cost for placing an order
$\mu(t)$: Deterioration rate of on-hand inventory during cycle.
Q: On-hand inventory at time $t = 0$
d: Cost of the purchasing items
h: Holding cost per unit

The following assumptions are made in the study:

i. $\mu(t) = \alpha\beta\, t^{\beta-1}$ is the deterioration rate, where α and β are the parameters and lie between $0 < \alpha < 1$, $0 < \beta < 1$.

ii. $D(t) = \left\{ a + bt,\ a,\ \alpha p^{-\beta} \right\}$, where a is the constant and b, α and β are the parameters.

iii. Shortages are not allowed.

iv. Demand rates are varying for the different stage of the model.

v. The inventory system involves only one stocking point; Q represents the maximum inventory level.

vi. The model is formulated for single item with no replenishment.

5.3   FORMULATION OF THE MATHEMATICAL MODEL

The inventory models available in the relevant literatures can be classified broadly on the basis of demand rates into the following three categories:

i. Model for inventory with linear demand rate
ii. Model for inventory with constant demand rates
iii. Model for inventory with selling price-dependent demand rates

Considerable amount of research work has been devoted on decaying inventory system. Different researcher developed many mathematical models from time to time in order to accommodate various realistic factors.

As the demand plays an important role for deteriorating inventory models, it can affect the profit, so many researchers recognised and studied the demand variations in graph from the view point of real-life situations. The behaviour of the inventory level at any time 't' is depicted in Figure 5.1. There may arise three cases according to the demand rates. The model starts with the initial level of the stock at time $t = 0$, and it reached to the maximum level at t_1. Now from time interval $[t_1, t_2]$, the demand rate is assumed as constant and between the time interval $[t_2, t_3]$ the demand is assumed to be selling price dependent due to the combined effect of demand deterioration, and the inventory stock is decaying which reached to zero at time $t = t_3$. The model is developed without the shortages, so at the end of the one cycle, the production again starts at the same time $t = t_3$. The graphical representation of the model is given in Figure 5.1.

The mathematical formulation of the model is calculated in the following three different intervals.

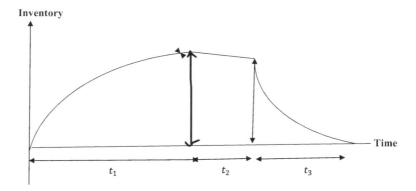

FIGURE 5.1   Graphical representation of the model.

Case 1: In this case, we have assumed the demand rate linear

$$D(t) = a + bt \text{ and } \mu(t) = \alpha\beta t^{\beta-1}$$

In this section, the inventory model starting with no shortages is studied. At the beginning of the cycle inventory the level of inventory is Q. Due to the demand and deterioration there is gradual fall in the inventory during the period.

The differential equations showing the fluctuations of inventory with time t are shown as below:

$$\frac{dI_1}{dt} = -\mu(t)I_1(t) - D(t) \ 0 \le t \le t_1 \tag{5.1}$$

$$\frac{dI_1}{dt} = -D(t) \ t_1 \le t \le T \tag{5.2}$$

Subject to the condition,

$$I_1(0) = Q, \ I_2(T) = 0$$

Due to the combined effect of demand and deterioration of the items it ultimately fall to zero at $t = T$. Hence, the differential equation representing the inventory level $[0, T]$ is given by

$$I_1(t) = Q - e^{-\alpha t^\beta}\left[at + \frac{bt^2}{2} + \frac{a\alpha t^{\beta+1}}{\beta+1} + b\alpha\frac{t^{\beta+2}}{\beta+2}\right] \tag{5.3}$$

$$I_2(t) = -\left[at + \frac{bt^2}{2}\right] \tag{5.4}$$

Now, to calculate the following cost:

i. Ordering cost:
   Ordering cost per unit time is given by

$$OC = \frac{A}{T} \tag{5.5}$$

ii. Deterioration cost:
   Deterioration cost per unit time is given by

$$DC = \frac{d}{T}\left[\int_0^{t_1} \alpha\beta t^{\beta-1} I_1(t)\,dt\right] \tag{5.6}$$

$$DC = \left[Q - \int_0^{t_1} D(t)d(t)\right] = \left[Q - \int_0^{t_1}(a+bt)dt\right]d$$

$$DC = \left[Q - at_1 - b\frac{t_1^2}{2}\right] \tag{5.7}$$

iii. Holding cost:
   The holding cost per unit time is given by

$$HC = h\int_0^{t_1} I_1(t)\,dt = h\int_0^{t_1}\left\{Q - e^{-\alpha t^\beta}\left[at + \frac{bt^2}{2} + \frac{a\alpha t^{\beta+1}}{\beta+1} + b\alpha\frac{t^{\beta+2}}{\beta+2}\right]\right\}dt$$

$$= h\left\{\int_0^{t_1} Q\,dt - a\frac{t_1^2}{2} - a\alpha\frac{t_1^{\beta+2}}{\beta+2} + b\frac{t_1^3}{6} - \alpha b\frac{t_1^{\beta+3}}{2(\beta+3)}\right\} \tag{5.8}$$

iv. Shortages cost:
   The shortages cost of the system is given by

$$SC = S\int_{t_1}^{T}(a+bt)\,dt$$

$$= S\left[a(T-t_1) + \frac{b}{2}\left(T^2 - t_1^2\right)\right] \tag{5.9}$$

v. Purchasing cost:
   The purchasing cost of the system is given by

$$PC = Qd$$

vi. Sales revenue cost:
Sales revenue cost per unit time is given by

$$SR = P\left[\int_0^{t_1}(a+bt)dt + \int_{t_1}^{T}(a+bt)dt\right]$$

$$SR = P\left[\left(at_1 + b\frac{t_1^2}{2}\right) + a(T - t_1) + b\left(\frac{T^2}{2} - \frac{t_1^2}{2}\right)\right] \quad (5.10)$$

Case 2: When $D(t) = a$, where a is the constant, $a > 0$

From Eqs. (5.1) and (5.2), we get the solution

$$I_1(t) = Q - e^{-\alpha t^\beta}\left[at + a\alpha\frac{t^{\beta+1}}{\beta+1}\right] \quad (5.11)$$

$$I_2(t) = -at \quad (5.12)$$

The following costs calculated are:
i. Ordering cost:
The ordering cost per unit time is given by

$$OC = \frac{A}{T} \quad (5.13)$$

ii. Sales revenue cost per:
The sales revenue cost per unit time is given by

$$SR = P\left[at_1 + a(T - t_1)\right] \quad (5.14)$$

iii. Deterioration:
Deterioration cost per unit time is given by

$$DC = \left[Q - \int_0^{t_1}D(t)d(t)\right]d = \left[Q - \int_0^{t_1}(a)dt\right]d$$

$$DC = \left[Q - at_1\right]d \quad (5.15)$$

iv. Holding cost:
The holding cost per unit time is given by

$$HC = h\int_0^{t_1}I_1(t)dt = h\int_0^{t_1}\left\{Q - e^{-\alpha t^\beta}\left[at + \frac{a\alpha t^{\beta+1}}{\beta+1}\right]\right\}dt$$

$$= h\left\{\int_0^{t_1} Q\ dt - a\frac{t_1^2}{2} - a\alpha\frac{t_1^{\beta+2}}{\beta+2}\right\} \tag{5.16}$$

v. Shortage's cost:
    The shortages cost of the system is given by

$$SC = S\int_{t_1}^{T}(a)dt$$

$$= S\left[a(T - t_1)\right] \tag{5.17}$$

vi. Purchasing cost:
    The purchasing cost of the system is given by

$$PC = Qd$$

Case 3: When demand is taken as selling price dependent, i.e. $D(t) = \dfrac{\alpha}{p^\beta}$, where α and β are parameters $0 < \alpha < 1$, $0 < \beta < 1$ and P is the selling price of the items.

Using Eqs. (5.1) and (5.2), the solution of the equation becomes

$$I_1(t) = Q - e^{-\alpha t^\beta}\frac{\alpha}{p^\beta}\left(t + \alpha\frac{t^{\beta+1}}{\beta+1}\right) \tag{5.18}$$

$$I_2(t) = -\alpha\frac{t}{p^\beta} \tag{5.19}$$

Now in order to find out the different cost of the system to calculate the following cost:

i. Ordering cost:
    Ordering cost per unit time is given by

$$OC = \frac{A}{T} \tag{5.20}$$

ii. Sales revenue cost:
    Sales revenue cost per unit time is given by

$$SR = P\left[\left(\int_0^{t_1}\alpha p^{-\beta}dt + \int_{t_1}^{T}\alpha p^{-\beta}\right)dt\right]$$

$$SR = P\left[\alpha p^{-\beta}(t_1) + \alpha p^{-\beta}(T - t_1)\right]$$

iii. Deterioration cost:
Deterioration cost per unit time is given by

$$DC = \left[Q - \int_0^{t_1} D(t) d(t) \right] d = \left[Q - \int_0^{t_1} \alpha p^{-\beta} dt \right] d$$

$$DC = \left[Q - \alpha p^{-\beta} t_1 \right] d \qquad (5.21)$$

iv. Holding cost:
The holding cost per unit time is given by

$$HC = h \int_0^{t_1} I_1(t) dt = h \int_0^{t_1} \left\{ Q - e^{-\alpha t^{\beta}} \alpha p^{-\beta} \left[t + \frac{\alpha t^{\beta+1}}{\beta+1} \right] \right\} dt$$

$$= h \left\{ \int_0^{t_1} Q \, dt - \alpha p^{-\beta} \frac{t_1^2}{2} - a\alpha \frac{t_1^{\beta+2}}{(\beta+1)(\beta+2)} \right\} \qquad (5.22)$$

v. Shortages cost:
The shortages cost of the system is given by

$$SC = S \int_{t_1}^{T} \alpha p^{-\beta} dt$$

$$= S\left[\alpha(T - t_1) \right] \qquad (5.23)$$

vi. Purchasing cost:
The purchasing cost of the system is given by

$$PC = Qd \qquad (5.24)$$

Therefore, the total profit per unit time of our model is obtained as follows:

$$P(t_1, t_2) = \frac{1}{T} \left\{ \begin{array}{l} \text{Sales revenue-purchase cost-ordering cost-holding cost} \\ \text{in RW-holding cost in OW-back order cost-opportunity cost} \end{array} \right\} \qquad (5.25)$$

To maximise the total average profit per unit time (P), the optimal values of t_1 and t_2 can be obtained by solving the following equations simultaneously:

$$\frac{\partial P}{\partial t_1} = 0, \text{ and } \frac{\partial P}{\partial t_2} = 0 \qquad (5.26)$$

provided, it satisfy the following conditions:

$$\frac{\partial^2 P}{\partial t_1^2} > 0, \frac{\partial^2 P}{\partial t_2^2} > 0 \qquad (5.27)$$

$$\text{and } \left(\frac{\partial^2 P}{\partial t_1^2}\right)\left(\frac{\partial^2 P}{\partial t_2^2}\right) - \left(\frac{\partial^2 P}{\partial t_1 \partial t_2}\right)^2 > 0 \qquad (5.28)$$

Equations (5.25) and (5.26) are highly non-linear and hence are solved with the help of mathematical software ***MATHEMATICA 5.2***. With the use of these optimal values, Eq. (5.20) provides the maximum total average profit per unit time of the system in consideration.

5.4 NUMERICAL EXAMPLES

Four numerical examples are used to illustrate all results in this chapter. The necessary parameters and the optimal solutions of the four examples are presented.

Example: $a = 800$, $b = 0.008$, $\theta_1 = 0.09$, $\theta_2 = 0.06$, $t_w = 20$, $T = 80$, $r = 0.06$, $S_0 = 100$, $q_r = 200$, $W = 600$, $A = 100$, $\delta = 0.7$, $c_1 = 0.6$, $c_2 = 0.3$, $c_3 = 0.5$, $c_4 = 5$, $c_5 = 12$, $s = 14$, $t_1 = 50.6455$, $t_2 == 67.9233$, $H_{RW} = 44.9501$, $H_{OW} = 2218.28$, $BC = 122889$, $OC = 47938$, $Q = 283353$, **Total profit (π)** $= 60440.5$.

5.5 SENSITIVITY ANALYSIS

Based on the values used in the above example in the model, the authors have examined the sensitivity analysis by changing some parameters one at a time and keeping the rest fixed.

The authors have calculated the sensitivity analysis based on different parameters. The outcome of the result is compared.

5.6 OBSERVATIONS

In the present study, the authors have calculated the sensitivity analysis based on the parameters used in the study. The authors have made changes in the parameters by −50%, −25%, 0%, 25% and 50%. Some important inferences drawn from Table 5.1 and Figures 5.2–5.5 are as follows:

i. Table No. 1 shows that as the value of A goes from −50% to +50%, the values of t_1, T, I_{max} and total cost also increase.
ii. Table No. 1 shows that as the value of a goes from −50% to +50%, the values of T and I_{max} decrease, while the values of t_1 and the total cost increase rapidly.
iii. Table No. 1 shows that as the value of b goes from −50% to +50%, the values of T and total cost increase, while the values of t_1, T and I_{max} decrease rapidly.
iv. Table No. 1 shows that as the value of p goes from −50% to +50%, the values of T and total cost increase, while the values of t_1, T and I_{max} decrease.

TABLE 5.1

Sensitivity Analysis w.r.t. Various Parameters

Parameter	%	Changed Value	t_1	T	I_{max}	TC
A	+50%	75	1.34725	2.69451	169.0449	359.918
	+25%	62.5	1.22741	2.40402	130.22971	354.026
	0	50	1.10003	1.20006	114.77403	349.788
	−25%	37.5	0.97816	1.98821	71.21354	341.498
	−50%	25	0.87699	1.68724	32.21354	333.805
a	+50	30	2.43291	1.09105	19.33721	489.125
	+25	25	1.22886	1.10554	53.97692	431.226
	0	20	1.10003	2.20006	114.77403	349.788
	−25	15	0.92214	2.31269	148.29284	266.394
	−50	10	0.78764	2.39991	187.76542	200.947
b	+50	1.8	1.09371	2.18739	90.53786	354.985
	+25	1.4	1.09685	2.19369	102.73451	352.063
	0	1.2	1.10003	2.20006	114.77403	339.788
	−25	0.8	1.10324	2.20647	123.95031	348.072
	−50	0.6	1.12009	2.22018	149.24731	346.857
P	+50	56	0.53185	1.59475	107.98073	361.688
	+25	50	0.86818	1.82913	110.76742	357.296
	0	40	1.10003	2.20006	114.77403	349.788
	−25	30	1.33422	2.62627	143.91819	335.358
	−50	20	1.67151	2.91326	201.21387	317.963

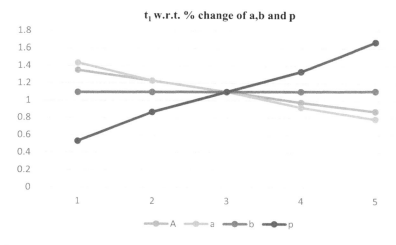

FIGURE 5.2 t_1 vs. change in parameters.

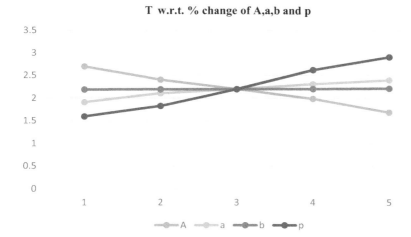

FIGURE 5.3 T vs. change in parameters.

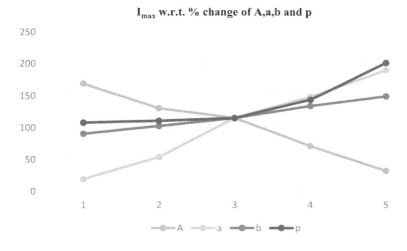

FIGURE 5.4 I_{max} vs. change in parameters.

5.7 CONCLUSIONS

In this present study, by considering a reliable business purpose model, we have taken different demands for the different stages of the model and developed an inventory model for deteriorating items as well. To consider the effect of the various parameters on the business, the model is developed under time-dependent deterioration, where shortage is not allowed. The net profit function has been maximised for optimality.

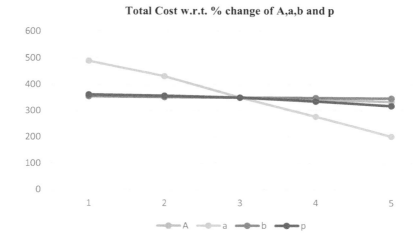

FIGURE 5.5   Total cost vs. change in parameters.

In this model, the effect of each cost and parameters are also investigated, where it has been proved that the net profit is maximised by making some changes in the parameters and their boundary conditions.

It is also obvious from the model that by assuming more variable-dependent demand, we can improve the profit, so the model is more important as per the market ups and downs.

The focus of the study was to control the shortages, and it is proved here that we can stop or minimise the shortages in the business by applying a multivariable-based or -dependent models.

It is also found that by considering different demands on different stages of model, it is more economical to maximise the profit. Further, we can develop this model for shortages, partial backlogging and for different permissible delay conditions.

5.8   FUTURE SCOPE OF THE STUDY

The present study may be more expanded by making some more changes in the main parameters considered under this study like the model can also be developed under shortage and trade credits parameters etc.

REFERENCES

1. Ghare PM. A model for an exponentially decaying inventory. *Journal of Industrial Engineering.* 1963;14:238–43.
2. Pakkala TP, Achary KK. Two level storage inventory model for deteriorating items with bulk release rule. *Opsearch.* 1994;31:215.
3. Goswami A, Chaudhuri KS. An economic order quantity model for items with two levels of storage for a linear trend in demand. *Journal of the Operational Research Society.* 1992;43(2):157–67.

4. Bhunia AK, Maiti M. A two warehouse inventory model for a linear trend in demand. *Opsearch*. 1994;31:318.
5. Benkherouf L. A deterministic order level inventory model for deteriorating items with two storage facilities. *International Journal of Production Economics*. 1997;48(2):167–75.
6. Kar S, Bhunia AK, Maiti M. Deterministic inventory model with two levels of storage, a linear trend in demand and a fixed time horizon. *Computers & Operations Research*. 2001;28(13):1315–31.
7. Zhou YW, Yang SL. A two-warehouse inventory model for items with stock-level-dependent demand rate. *International Journal of Production Economics*. 2005;95(2):215–28.
8. Wee HM, Yu JC, Law ST. Two-warehouse inventory model with partial backordering and Weibull distribution deterioration under inflation. *Journal of the Chinese Institute of Industrial Engineers*. 2005;22(6):451–62.
9. Hsieh TP, Dye CY, Ouyang LY. Determining optimal lot size for a two-warehouse system with deterioration and shortages using net present value. *European Journal of Operational Research*. 2008;191(1):182–92.
10. Ghosh S, Chakrabarty T. An order-level inventory model under two level storage system with time-dependent demand. *Opsearch*. 2009;46(3):335–44.
11. Jaggi CK, Verma P. Two-warehouse inventory model for deteriorating items with linear trend in demand and shortages under inflationary conditions. *International Journal of Procurement Management*. 2010;3(1):54–71.
12. Sharma S, Singh SR. An inventory model for decaying items, considering multi variate consumption rate with partial backlogging. *Indian Journal of Science and Technology*. 2013;6(7):4870–80.
13. Sharma S, Singh S, Dem H. Two-warehouse production policy for different demands under volume flexibility. *International Journal of Industrial Engineering Computations*. 2013;4(4):599–609.
14. Sharma S, Singh SR. EPQ model of deteriorating inventory with exponential demand rate under limited storage. *ADRRI Journal (Multidisciplinary)*. 2013;1(1):23–31.
15. Khanna A, Pritam P, Jaggi CK. Optimizing preservation strategies for deteriorating items with time-varying holding cost and stock-dependent demand. *Yugoslav Journal of Operations Research*. 2020;30(2):237–50.
16. Chauhan A, Tayal S. An order quantity scheme for ramp type demand and backlogging during stock out with discount strategy. *International Journal of Services Operations and Informatics*. 2021;11(1):27–40.
17. Rout C, Chakraborty D, Goswami A. A production inventory model for deteriorating items with backlog-dependent demand. *RAIRO: Recherche Opérationnelle*. 2021;55:549.
18. Saxena AK. Review on EOQ models for instantaneous and non-instantaneous deteriorating items. In *2021 10th International Conference on System Modeling & Advancement in Research Trends (SMART)*. 2021, pp. 312–315. IEEE.
19. Singh SR, Rani M. An EPQ model with life-time items with multivariate demand with markdown policy under shortages and inflation. In *Journal of Physics: Conference Series*, 2021 (Vol. 1854, No. 1, p. 012045). IOP Publishing.
20. Mohan V, Ota M, Kumar A. An inventory model for decaying items with Pareto distribution, time-dependent demand and shortages. *International Journal of Mathematics in Operational Research*. 2022;21(1):83–103.

6 Risk Analysis of an Alternator in a Co-generation Power Plant Utilising Intuitionistic Fuzzy Concept

Dinesh Kumar Kushwaha and Anish Sachdeva
Dr. B. R. Ambedkar National Institute of Technology

Dilbagh Panchal
National Institute of Technology Kurukshetra

CONTENTS

6.1 INTRODUCTION

Failure mode effect analysis (FMEA) is a widely accepted and applied tool for carrying out risk assessment, and therefore adopted by many researchers with modification of the approach and advancement of time [1]. FMEA is a reliability investigation technique effective to classify the failures affecting the functioning of other system/subsystem/components and requires the corrective steps to be taken in advance of proactive failure [2]. FMEA approach is also applied in the identification of failure modes related to other fields such as product, process and services [3]. Usually, risk assessment was conducted with the implementation of FMEA approach, characterised by computation of the RPN score magnitude relevant to various failure modes/failure causes. Risk priority number (RPN) scores for failure modes/failure causes are calculated by multiplication of

DOI: 10.1201/9781003140092-6

three risk factors. In past, traditional FMEA was applied by a band of researchers [4–6]. However, in practice, the FMEA approach has many drawbacks that have been widely pointed out [3]. FMEA generates the same magnitude of RPN score; as a consequence, the listed failure causes are all given the same priority, making the risk prioritisation a very difficult task. Similarity in RPN score values creates a state of confusion for making decisions related to prioritising and ranking of failure modes/causes of any industrial system. In such conditions, a wrong decision is generally made by the system analysts, and thereby the critical failure modes/causes are not diagnosed correctly.

Moreover, a comparative importance of weight for three risk factors was also not considered in calculating RPN score output values. As a result, an uncertainty element in ranking result also persists due to equal importance for three risk factors [7]. Also, traditional FMEA approach is also confined to crisp set theory concept and that is why, vagueness of the raw data obtained from the feedback of experts was also not considered, which results in vague outcomes.

The uncertainties/vagueness in the expert's opinion are considered by a fuzzy-set-theory-based approach. Fuzzy set is characterised by an element and its membership function, which lie between closed interval [0,1]. FMEA based on fuzzy set theory has been extensively applied by many researchers to consider uncertainties/vagueness in the data obtained by experts [8–11]. Fuzzy FMEA approach was applied to consider the risk factor in a hospital service industry and specifically in a sterilisation unit of it [12]. FMEA and Z-number theory-based Z-MOORA techniques were applied to carry the risk analysis of an automotive spare parts company [13]. Grey theory-based FMEA approach was proposed to consider the uncertainties and to pinpoint the critical failure causes associated with an electronics company [14]. But hesitation in expert knowledge was missing in fuzzy-set-theory-based FMEA approach, the limitation of which was overcome by intuitionistic fuzzy set (IFS) theory.

To consider the indeterminacy or hesitation in the feedback of experts in their domain of knowledge, IFS theory proposed by Atanassov in the year 1986, which takes into account the hesitation effect in experts' judgement by taking non-membership function and degree of indeterminacy. The application of intuitionistic fuzzy (IF) modelling-based hybrid multi-criteria decision making (MCDM) techniques is widely increased, capturing different areas of applications. Hybrid interval-valued MCDM approach was proposed to study the risk analysis for hospital service applying IF concept [15]. Intuitionistic fuzzy weighing average (IFWA) operator-based IF-Technique for Order of Preference by Similarity to Ideal Solution (IF-TOPSIS) was applied to find hazardous cause in a gas refinery plant [16].. An integrated IF-FMEA approach-based framework was presented applying VlseKriterijumska Optimizacija I Kompromisno Resenje approach for risk analysis in ship building industry [18].

FMEA approach to identify and screen out no-risk and low-risk components of the pump drilling system was presented by [19]. Two approaches, i.e. FMEA and fault tree analysis (FTA), were proposed for reliably identifying the failure criticality of modes connected with an additive manufacturing system in order to improve operating sustainability [20]. Reliability assessment of a waste water treatment plant, combining the concepts of event tree analysis (ETA) and FTA with fuzzy logic, was proposed to address the unit's long-term operation and pollution-related difficulties [21]. A hybrid fuzzy FMEA approach was proposed for risk prioritising involved in a aircraft landing system [22]. The use of a fuzzy-based approach to analyse risk in the photovoltaic power industry in China was discussed to determine operational sustainability [23]. Due to versatility, FMEA has been applied by various researchers [24–27].

From the literature, it has been concluded that the risk assessment of alternator of co-generation power plant has not been done. So, considering this as the gap of literature, we propose risk analysis of an alternator unit to identify all the failure causes responsible for frequent tripping of the considered system.

6.2 PRELIMINARIES OF IFS

Definition 1: An IFS in a universal set X is given by Eq. (6.1) as

$$\bar{A} = \left\{ \langle x, \mu_{\bar{A}}(x), \vartheta_{\bar{x}}(x) \rangle \mid x \in X \right\} \tag{6.1}$$

where $\mu_{\bar{A}}(x)$ is the membership function and $\vartheta_{\bar{A}}(x)$ is the non-membership function calculated by Eqs. (6.2) and (6.3), respectively:

$$\mu_{\bar{A}} : S \to [01], x \in X \to \mu_{\bar{A}}(x) \to [01] \tag{6.2}$$

$$\vartheta_{\bar{A}} : S \to [01], x \in X \to \vartheta_{\bar{A}}(x) \to [01] \tag{6.3}$$

The above two equations must satisfy Eq. (6.4) for an IFS:

$$\mu_{\bar{A}}(x) + \vartheta_{\bar{A}}(x) \leq 1 \text{ for all } x \in X \tag{6.4}$$

Definition 2: Degree of indeterminacy/hesitation is defined by Eq. (6.5) and given as

$$\pi_{\bar{A}}(x) = 1 - \mu_{\bar{A}}(x) + \vartheta_{\bar{A}}(x) \text{ for all } x \in X$$

6.3 IF-FMEA APPROACH

The steps applied in IF-FMEA approach are as follows:

Step 1: Assign the linguistic variables for the risk factor against intuitionistic fuzzy number (IFN) as shown in Table 6.1.
Step 2: Using Eq. (6.5), aggregation of the team members subjective opinions was done and given in Table 6.2.

$$\text{IFWA}(\alpha_2, \alpha_2, \ldots, \alpha_n) = \psi_1 \cdot \alpha_1 + \psi_2 \cdot \alpha_2 + \ldots + \psi_{1n} \cdot \alpha_n$$

$$= \left(1 - \prod_{i=1}^{n} \left(1 - \mu_{\alpha_i} \right)^{wi}, \prod_{i=1}^{n} \left(v_{\alpha_i} \right)^{wi} \right) \tag{6.5}$$

where $\left(\psi_{11}, \psi_{12}, \ldots \psi_{1n} \right)^{T}$ is called the weight vector of α_i $(i = 1, 2, 3, \ldots n)$, with $\psi_{1i} \in [0\ 1]$ and $\sum_{i=1}^{n} \psi_{1i.} = 1$.

TABLE 6.1
**Linguistic Term for Rating the
Failure Causes**

Linguistic Variables	IFNs
Very low	(0.25, 0.70)
Low	(0.30, 0.60)
Medium low	(0.40, 0.50)
Medium	(0.50, 0.50)
Medium high	(0.60, 0.30)
High	(0.70, 0.20)
Very high	(0.75, 0.20)

TABLE 6.2
**Failure Modes Assessment in Terms of Feedback for Risk Factors from Three
Experts**

Failure Causes	Occurrence (O)			Severity (S)			Detection (D)		
	TM1	TM2	TM3	TM1	TM2	TM3	TM1	TM2	TM3
RT_1	M	L	VL	M	M	H	C	C	R
RT_2	MH	VL	L	M	MH	L	VH	C	C
RT_3	M	M	MH	VH	H	M	VH	VH	C
OV_4	H	M	H	M	L	VL	C	VH	H
OV_5	H	VH	M	H	MH	H	L	C	VH
OV_6	M	ML	L	MH	H	H	M	MH	H
OV_7	L	MH	M	H	VH	VH	H	VH	C
PP_8	L	VL	MH	VH	H	M	VH	VH	C
PP_9	M	VL	VH	VH	VH	VH	VH	VH	C
PP_{10}	M	M	M	H	H	VH	C	C	M
PP_{11}	VH	VH	MH	VH	VH	VH	C	C	C
PP_{12}	M	M	M	H	VH	VH	C	C	M

Step 3: Establish reference series for risk factors O, S and D.
Step 4: Tabulate the IF-RPN score for all listed failure causes using Eq. (6.6).

$$\text{IF-RPN} = dIFD(O) \times dIFD(S) \times dIFD(D) \qquad (6.6)$$

Step 5: Rank all the failure causes in descending order against the value of
IF-RPN output score as obtained.

6.4   INDUSTRIAL CASE STUDY

The proposed methodology is represented with an application of IF-FMEA approach
to carry risk assessment of an alternator in a co-generation power plant of sugar

industry located in the western part of Uttar Pradesh. Alternator is the most crucial subsystem of any sugar industry, as it generates power to fulfil the power demand of other subsystems/components of sugar mill as well as fulfil the residential load requirement. Alternator is an integrated subsystem of co-generation plant and comprises rotor, oil tank, pumps and valves arranged in series/parallel. The function of an alternator is to generate power. If an interruption in the operation of alternator is observed, it will affect all the other subsystems/components of sugar mill and bring the mill to halt, resulting in a major bottleneck in the production so as to continuously run the mill for 24/7, and the subsystem/equipment/components of alternator unit is also highly important. To upkeep the high availability, failure risk associated with various subsystem/equipment/components is very essential in a sugar mill. A sketch of alternator unit is shown in Figure 6.1.

6.4.1 APPLICATION OF PROPOSED METHODOLOGY

Select the FMEA team members to obtain their feedback associated with the list of various failure causes of different components of the alternator unit. The team members identify the 12 failure causes, and the risk factors are evaluated by these experts separately as shown in the developed FMEA sheet (Table 6.3).

The IFNs given in Table 6.1 and linguistic variables in Table 6.2 are used to aggregate the values of three team members' opinion with the relative weight assigned are 0.30, 0.45 and 0.25. The team members were qualified and experienced persons who were involved in the operation and maintenance of the alternator. The aggregated feedback obtained was calculated by using Eq. (6.5) and is given in Table 6.4.

After calculating the aggregated IFNs, the distance between two IFNs was calculated by establishing the reference series. $\bar{A}_0 = \left[(0,1),(0,1)\ldots(0,1)\right]$ and the results (IF-RPN) were tabulated by applying Eq. (6.6). Finally, IF-RPN score values and the ranks of failure causes are presented in Table 6.5.

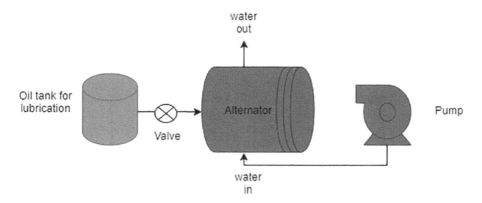

FIGURE 6.1 Alternator unit.

TABLE 6.3
FMEA Sheet of Listed Failure Causes

Subsystem/ Components	Function	Mode of Potential Failure	Effect of Potential Failure	Cause of Potential Failure
Rotor	To produce the required torque	Loud vibrations from bearings	Hamper the operation of alternator	Mechanical fault (RT1)
				Poor bearing lubrication (RT2)
Armature winding	To generate emf in armature winding	Deterioration of winding	Low emf generation	High temperature (RT3)
Oil tank and valves	Supply lubricating oil to bearings	Leakage	Loss of supply of oil	Poor welding (OV4)
				Low grade of material (OV5)
				Gland leakage (OV6)
				Seat rupture (OV7)
Pump	To supply water for cooling the windings	Leakage	Poor cooling of winding	Mechanical and electrical fault (PP8)
				Impeller failure (PP9)
				Bearing housing failure (PP10)
				Gasket rupture (PP11)
				Overheating of motor (PP12)

TABLE 6.4
Aggregated IFNs under O, S and D

S. No.	Failure cause	O		S		D	
1	RT_1	0.3663	0.4053	0.4824	0.5883	0.6984	0.8175
2	RT_2	0.4184	0.5114	0.4336	0.5352	0.7451	0.8519
3	RT_3	0.5105	0.5444	0.6478	0.7274	0.7121	0.8177
4	OV_4	0.6286	0.7147	0.3184	0.4053	0.7184	0.8182
5	OV_5	0.6573	0.7274	0.6527	0.7552	0.6277	0.7460
6	OV_6	0.4053	0.4582	0.6425	0.7451	0.5572	0.5739
7	OV_7	0.4461	0.5223	0.7118	0.7832	0.6938	0.7856
8	PP_8	0.3675	0.4579	0.6478	0.7274	0.7121	0.8177
9	PP_9	0.5084	0.5446	0.7321	0.7832	0.7121	0.8177
10	PP_{10}	0.4824	0.4824	0.6956	0.7832	0.7274	0.8322
11	PP_{11}	0.6986	0.7601	0.7321	0.7832	0.7832	0.8878
12	PP_{12}	0.4824	0.4824	0.7118	0.7832	0.7274	0.8322

TABLE 6.5
IFD and RPN Score Values of All the Failure Causes

Failure Cause	O	S	D	IF-RPN	Rank
RT_1	0.6637	0.6882	0.7896	0.36066	12
RT_2	0.6627	0.6660	0.8191	0.36149	11
RT_3	0.7382	0.7840	0.8032	0.46492	5
OV_4	0.7713	0.6157	0.8093	0.38435	10
OV_5	0.7935	0.7751	0.7547	0.46420	6
OV_6	0.6762	0.7700	0.7702	0.40101	9
OV_7	0.6850	0.8202	0.8010	0.44998	7
PP_8	0.6385	0.7840	0.8032	0.40212	8
PP_9	0.7361	0.8404	0.8032	0.49693	2
PP_{10}	0.7412	0.8040	0.8113	0.48346	4
PP_{11}	0.8186	0.8404	0.8393	0.57744	1
PP_{12}	0.7412	0.8202	0.8113	0.49320	3

6.5 RESULT AND DISCUSSIONS

From Table 6.5, it is revealed that failure cause gasket rupture (PP_{11}) with IF-RPN score 0.57744, associated with the pump, which is employed for cooling the armature winding, has been detected as the most critical failure cause with rank first, so the first priority should be given to upkeep the rupture of gasket to avoid the burnout or damage of armature windings. Failure cause (RT_1) of rotor is ranked the twelfth one, and it is the last failure cause, so this failure cause should be given the least priority while framing maintenance policy, with IF-RPN score 0.36066. The IF-RPN score of two failure causes are not same which ensures easy prioritising and do not create any confusion. If the hesitation effect is taken into account in the FMEA, it is a beneficial and an effective way to propose for risk analysis of other repairable industrial systems. Apart from that, there are less risky failure causes too. PP_9, PP_{12} and PP_{10} are ranked in the order of second, third and fourth with their corresponding IF-RPN score values −0.49693, 0.49320 and 0.48346 in terms of prioritising all the failure causes relevant to pump of the alternator unit of the considered sugar mill plant. The other ranking of the failure causes is shown in Table 6.5. A time-based maintenance scheme considering the risk related to the rank of critical failure causes will be beneficial for maintenance manager to ensure increase in availability and in upkeep of the alternator unit.

6.6 CONCLUSIONS

The risk evaluation of the alternator unit was illustrated using an IF-FMEA approach based on intuitionistic fuzzy concepts. The benefits of the aforementioned technique come from the fact that it considers membership, non-membership and the degree of hesitation in an expert's knowledge horizon. Another benefit of this approach was the aggregate of expert subjective feedback, which was done by IFWA in order to take into account the average effect of numerous expert viewpoints. The IF-RPN output score and related ranks of the considered system were computed, allowing the maintenance manager to identify the critical component and failure reasons of the alternator unit system and, as a result, to conduct risk analysis. The long-run availability and maintainability of the considered unit can be only ensured by framing a correct maintenance policy based on the results obtained by the risk analysis. In other words, advanced types of maintenance are recommended for the highly riskier failure causes and the components relevant to that failure cause. In addition to this, those failure causes fall in the category of less risky one, and a simple maintenance scheme can be recommended. In essence, different maintenance schemes are endorsed for different failure causes and its associated party depending on the level of risk and the severity of damage it can cause to production, personnel and environment. Apart from those advantages, the current approach can be extended by considering the objective weight of the risk factor, which is not taken into account in this study. This will be a scope for the future work to be considered. Also, this work can be extended to other subsystems of process industry for carrying out the risk analysis under intuitionistic fuzzy environment.

REFERENCES

1. Panchal D, Singh AK, Chatterjee P, Zavadskas EK, Keshavarz-Ghorabaee M. A new fuzzy methodology-based structured framework for RAM and risk analysis. *Applied Soft Computing.* 2019;74:242–54.
2. Bowles JB, Peláez CE. Fuzzy logic prioritization of failures in a system failure mode, effects and criticality analysis. *Reliability Engineering & System Safety.* 1995;50(2):203–13.
3. Wang YM, Chin KS, Poon GK, Yang JB. Risk evaluation in failure mode and effects analysis using fuzzy weighted geometric mean. *Expert Systems with Applications.* 2009;36(2):1195–207.
4. Nie RX, Tian ZP, Wang XK, Wang JQ, Wang TL. Risk evaluation by FMEA of supercritical water gasification system using multi-granular linguistic distribution assessment. *Knowledge-Based Systems.* 2018;162:185–201.
5. Fattahi R, Khalilzadeh M. Risk evaluation using a novel hybrid method based on FMEA, extended MULTIMOORA, and AHP methods under fuzzy environment. *Safety Science.* 2018;102:290–300.
6. Li Z, Chen L. A novel evidential FMEA method by integrating fuzzy belief structure and grey relational projection method. *Engineering Applications of Artificial Intelligence.* 2019;77:136–47.
7. Liu HC, Liu L, Lin QL. Fuzzy failure mode and effects analysis using fuzzy evidential reasoning and belief rule-based methodology. *IEEE Transactions on Reliability.* 2013;62(1):23–36.
8. Feili HR, Akar N, Lotfizadeh H, Bairampour M, Nasiri S. Risk analysis of geothermal power plants using Failure Modes and Effects Analysis (FMEA) technique. *Energy Conversion and Management.* 2013;72:69–76.
9. Villarini M, Cesarotti V, Alfonsi L, Introna V. Optimization of photovoltaic maintenance plan by means of a FMEA approach based on real data. *Energy Conversion and Management.* 2017;152:1–2.
10. Panchal D, Kumar D. Stochastic behaviour analysis of real industrial system. *International Journal of System Assurance Engineering and Management.* 2017;8(2):1126–42.

11. Panchal D, Jamwal U, Srivastava P, Kamboj K, Sharma R. Fuzzy methodology application for failure analysis of transmission system. *International Journal of Mathematics in Operational Research*. 2018;12(2):220–37.
12. Dağsuyu C, Göçmen E, Narlı M, Kokangül A. Classical and fuzzy FMEA risk analysis in a sterilization unit. *Computers & Industrial Engineering*. 2016;101:286–94.
13. Ghoushchi SJ, Yousefi S, Khazaeili M. An extended FMEA approach based on the Z-MOORA and fuzzy BWM for prioritization of failures. *Applied Soft Computing*. 2019;81:105505.
14. Lo HW, Liou JJ. A novel multiple-criteria decision-making-based FMEA model for risk assessment. *Applied Soft Computing*. 2018;73:684–96.
15. Wang LE, Liu HC, Quan MY. Evaluating the risk of failure modes with a hybrid MCDM model under interval-valued intuitionistic fuzzy environments. *Computers & Industrial Engineering*. 2016;102:175–85.
16. Yazdi M. Risk assessment based on novel intuitionistic fuzzy-hybrid-modified TOPSIS approach. *Safety Science*. 2018;110:438–48.
17. Sayyadi Tooranloo H, Ayatollah AS, Alboghobish S. Evaluating knowledge management failure factors using intuitionistic fuzzy FMEA approach. *Knowledge and Information Systems*. 2018;57(1):183–205.
18. Efe B. Analysis of operational safety risks in shipbuilding using failure mode and effect analysis approach. *Ocean Engineering*. 2019;187:106214.
19. Tang Y, Liu Q, Jing J, Yang Y, Zou Z. A framework for identification of maintenance significant items in reliability centered maintenance. *Energy*. 2017;118:1295–303.
20. Peeters JF, Basten RJ, Tinga T. Improving failure analysis efficiency by combining FTA and FMEA in a recursive manner. *Reliability Engineering & System Safety*. 2018;172:36–44.
21. Piadeh F, Ahmadi M, Behzadian K. Reliability assessment for hybrid systems of advanced treatment units of industrial wastewater reuse using combined event tree and fuzzy fault tree analyses. *Journal of Cleaner Production*. 2018;201:958–73.
22. Wang W, Liu X, Qin Y, Fu Y. A risk evaluation and prioritization method for FMEA with prospect theory and Choquet integral. *Safety Science*. 2018;110:152–63.
23. Wu Y, Li L, Song Z, Lin X. Risk assessment on offshore photovoltaic power generation projects in China based on a fuzzy analysis framework. *Journal of Cleaner Production*. 2019;215:46–62.
24. Kushwaha DK, Panchal D, Sachdeva A. Risk analysis of cutting system under intuitionistic fuzzy environment. *Reports in Mechanical Engineering*. 2020;1(1):162–73.
25. Panchal D, Kumar D. Risk analysis of compressor house unit in thermal power plant using integrated fuzzy FMEA and GRA approach. *International Journal of Industrial and Systems Engineering*. 2017;25(2):228–50.
26. Panchal D, Kumar D. Integrated framework for behaviour analysis in a process plant. *Journal of Loss Prevention in the Process Industries*. 2016;40:147–61.
27. Gopal N, Panchal D. A structured framework for performance optimization using JBLTO, FCOPRAS and FCODAS methodologies. *Journal of Quality in Maintenance Engineering*. 2022. DOI 10.1108/JQME-11-2021-0087.

7 Risk Analysis of Cheese Packaging Machine Using FMEA and FCODAS Approach

Nand Gopal
Dr. B.R. Ambedkar National Institute of Technology

Dilbagh Panchal
National Institute of Technology Kurukshetra

CONTENTS

7.1   INTRODUCTION

Risk analysis has been seen as a critical necessity for the heavy process industry's long-term viability. It is important in all aspects of the manufacturing process, particularly during the product development cycle, assembly and packaging creation. It becomes more and more crucial when the industries producing perishable products such as cheese, curd and other dairy products and that too have to deliver timely. Due to the perishable nature of milk, it is critical to process it as soon as possible. Furthermore, the operating/operational costs of milk plants are extremely high, and the plant's frequent breakdown necessitates plant repair and maintenance. To avoid the above-discussed situation, all equipment must be in operational state for 24×7 in order to achieve optimal productivity, i.e. consistent service with maximum performance. The end result of the produced dairy product totally depends on the proper packaging and for the same,

DOI: 10.1201/9781003140092-7

packaging machine must work round the clock. Otherwise, breakdown/improper functioning of packaging machine will cause halt the process for which product has to wait until it comes into its operational condition with full capacity. Hence, study of risk issues associated with various components of cheese packaging machine (CPM) is must.

7.2 LITERATURE REVIEW

Many researchers have already conducted risk-based analysis for many process industries such as paper mill [1], sugar mill [2], power plant [3], casting industries [4] and chemical industry [5]. Various studies have used failure mode effect analysis (FMEA) and decision-making methodologies-based risk framework in recent years. Sharma and Sharma [6] expounded a new FMEA model based on a fuzzy inference system (FIS) for analysing risk issues in a paper industry. Panchal and Kumar [7] described the use of a fuzzy rule-based FMEA risk framework to do risk analysis of a coal-fired thermal power plant's water treatment plant. Panchal et al. [8] developed a novel risk-based integrated model using the FMEA approach and the grey relation analysis (GRA) technology. The collected results were compared to techniques, and the crucial components of the urea fertiliser industry were recognised as a result. Kushwaha et al. [9] presented a new intuitionistic fuzzy-based FMEA model. Several researchers have used many Multi-Criteria Decision-Making (MCDM) approaches in different fields. Chatterjee and Chakraborty [10] proposed a meta-model for determining cotton fibre's technological value. Kushwaha et al. [9] proposed intuitionistic fuzzy Technique for Order of Preference by Similarity to Ideal Solution (TOPSIS), which was used to compute the ranking of failure causes in sugar mill industry's cutting system unit. Agarwal et al. [11] found a contextual relationship between the green supply chain management barriers through Interpretive Structural Modelling (ISM) and Matrice d'Impacts Croisés Multiplication Appliquée á un Classement (MICMAC) analysis. Kumar et al. [12] used Analytical Hierarchy Process (AHP)-fuzzy TOPSIS-based approach to evaluate the performance of cold supply chain. Panchal et al. [13] expounded three-phase MCDM approaches, i.e. fuzzy AHP, fuzzy TOPSIS and fuzzy Evaluation based on Distance from Average Solution (EDAS) to select sustainable oil for foundry industries. Kumar et al. [14] proposed AHP-TOPSIS-based framework to analyse various critical criteria of vaccine cold supply chain for a pharmaceutical industry. These MCDM techniques have been shown to be highly beneficial in solving various decision-related problems [15].

From the above-studied literature, it has been observed that the application of integrated FMEA and FCODAS for analysing the risk issues of CPM has not yet been reported. Considering this as the gap, the application of both these integrated approaches is presented on CPM in a milk process industry located in Punjab province of India.

7.3 BASIC CONCEPT OF FUZZY SET THEORY

7.3.1 CRISP AND FUZZY SET

Crisp logic is concerned with absolutes, true or false, with no middle ground. For example, if the temperature exceeds 75°F, it is hot; otherwise, it is not hot.

Zadeh proposed the fuzzy set theory in 1965. The theory of fuzzy set is defined by its membership functions. The degree of belonging is expressed by the membership function, which is a function in [0,1] [14]. If the number (in the range [0,1]) is greater, then the degree of belonging will also be greater. For any fuzzy set $\tilde{F}$, the function $\mu_{\tilde{F}}$ represents the membership function for which $\mu_{\tilde{F}}(x)$ denotes the degree of membership that x, of the universal set X, belongs to set $\tilde{F}$ and is commonly written as a number between 0 and 1. It can be expressed by Eq. (7.1) as follows:

$$\mu_{\tilde{F}}(x): X \to [0,1] \tag{7.1}$$

7.3.2 MEMBERSHIP FUNCTION (MF)

It is a function that specifies the degree to which a given input belongs to a set. Various types of MF, such as triangular, piecewise linear, trapezoidal, Gaussian and singleton, have been used by various researchers to consider the vagueness in the collected details. Triangular membership function (TMF) has been used by many researchers to consider ambiguity in data for carrying out risk analysis for real-world engineering systems [2,5,16–19]. A TMF is mathematically defined by Eq. (7.2) as follows:

$$\mu_{\widetilde{A_{js}}}(x) = \begin{cases} \dfrac{x - L_{js}}{M_{js} - L_{js}}, & L_{js} \leq x \leq M_{js} \\ 1, & x = M_{js} \\ \dfrac{N_{js} - x}{N_{js} - M_{js}}, & M_{js} \leq x \leq N_{js} \\ 0, & \text{Otherwise} \end{cases} \tag{7.2}$$

7.3.3 FMEA APPROACH

FMEA analysis was first appeared in 1960, when NASA utilised it during the Apollo mission, a sort of space programme. It is a widely used method for determining the causes, effects and modes of failure in subsystems/components of a complex real-world industrial system [3,6,13,18,20–23]. The conventional FMEA approach is highly good for efficiently and effectively listing all of the qualitative information for the system. On the other hand, FMEA does not have any mathematical background for calculating risk priority number (RPN) $(O_f \times S \times O_d = \text{RPN})$, which requires the least effort for computing priority outcomes [24]. FMEA approach has been complained by several researchers [5,8,9,25,26] because of its inadequacies such as same RPN score for different failure causes, no exact base for RPN formula and three risk factors have equal weightage. As a result, using a decision approach like FCODAS improves the accuracy of the decision findings.

7.3.4 FCODAS APPROACH

FCODAS is a powerful approach that has been used by a number of researchers to solve complicated decision problems in a variety of fields [4,27–29]. When establishing the criticality of a collection of failure causes described using FMEA technique, CODAS examines the relative relevance of the distances. In addition to this, introducing a fuzzy concept into this technique, deftness can be improved. The steps of the FCODAS approach are as follows:

Step 1: Develop an initial fuzzy decision matrix $\left(\widetilde{DV}\right)$ for three risk factors represented as follows:

$$\widetilde{DV} = \widetilde{dv_{mn}} = \begin{bmatrix} \widetilde{dv}_{e11}, \widetilde{dv}_{f11}, \widetilde{dv}_{g11} & \cdots & \widetilde{dv}_{e13}, \widetilde{dv}_{f13}, \widetilde{dv}_{g13} \\ \widetilde{dv}_{e21}, \widetilde{dv}_{f21}, \widetilde{dv}_{g21} & \cdots & \widetilde{dv}_{e23}, \widetilde{dv}_{f23}, \widetilde{dv}_{g23} \\ \vdots & \vdots & \vdots \\ \widetilde{dv}_{ep1}, \widetilde{dv}_{fp1}, \widetilde{dv}_{gp1} & \cdots & \widetilde{dv}_{ep3}, \widetilde{dv}_{fp3}, \widetilde{dv}_{gp3} \end{bmatrix} \quad (7.3)$$

$; m = 1, \ldots p, \ \& \ n = 1 \text{ to } 3$

Step 2: Develop an average fuzzy decision matrix using Eq. (7.4).

$$\overline{dv}_{mn} = \frac{\left(\widetilde{dv}_{emn} + \widetilde{dv}_{fmn} + \widetilde{dv}_{gmn}\right)}{3} \quad (7.4)$$

where $\overline{dv}_{mn} = \left[\widetilde{dv}_{mn}^{lb}, \widetilde{dv}_{mn}^{mb}, \widetilde{dv}_{mn}^{ub}\right]$ implies a triangular fuzzy number (TFN); *lbo*, *mbo* and *ubo* signify lower, middle and upper bound.

Step 3: Convert fuzzified values into defuzzified (crisp) values using Eq. (7.5):

$$dv_{mn} = \frac{\left(\widetilde{dv}_{mn}^{lb} + 4\widetilde{dv}_{mn}^{mb} + \widetilde{dv}_{mn}^{ub}\right)}{6} \quad (7.5)$$

and develop decision matrix using defuzzified (crisp) values represented as follows:

$$dv = dv_{mn} = \begin{bmatrix} dv_{11} & dv_{12} & dv_{13} \\ dv_{21} & dv_{22} & dv_{23} \\ \vdots & \vdots & \vdots \\ dv_{p1} & dv_{p2} & dv_{p3} \end{bmatrix} \text{ where } m = 1, \ldots p, \text{ and } n = 1 \text{ to } 3 \quad (7.6)$$

Step 4: Create normalised beneficial and non-beneficial decision matrix as follows:

$$\overline{dv}_{mn} = \begin{cases} \dfrac{dv_{mn}}{\max\limits_m dv_{mn}} & \text{if } n \in bn_n \\[4mm] \dfrac{\min\limits_m dv_{mn}}{dv_{mn}} & \text{if } n \in nbn_{nb} \end{cases} \quad (7.7)$$

where bn_n and nbn_{nb} are beneficial and non-beneficial failure causes, respectively.

Step 5: Using Eq. (7.8), determine weighted matrix.

$$\widehat{dv}_{mn} = w_s \times \overline{dv}_{mn} \text{ where } m = 1,\ldots p, \text{ and } n = 1 \text{ to } 3 \tag{7.8}$$

where w_s represents weights of risk factors, and established normalised weighted matrix represented as follows:

$$\widehat{dv} = \widehat{dv}_{mn} = \begin{bmatrix} \widehat{dv}_{11} & \widehat{dv}_{12} & \widehat{dv}_{13} \\ \widehat{dv}_{21} & \widehat{dv}_{22} & \widehat{dv}_{23} \\ \vdots & \vdots & \vdots \\ \widehat{dv}_{p1} & \widehat{dv}_{p2} & \widehat{dv}_{p3} \end{bmatrix} \text{ where } m = 1,\ldots p, \text{ and } n = 1 \text{ to } 3 \tag{7.9}$$

Step 6: Tabulate negative-ideal solution $(\dot{\partial})$ matrix using Eq. (7.10).

$$\dot{\exists} = [\exists_n]_{1 \times 3} \tag{7.10}$$

where $\exists_n = {}^{min}_m \widehat{dv}_{mn}$.

Step 7: Calculate Euclidean (Eu_i) and Taxicab (Ta_i) distances as per Eqs. (7.11) and (7.12).

$$(Eu_i) = \sum_{n=1}^{3} Eu_i(\widehat{dv}_{mn}, \exists_n) \tag{7.11}$$

$$(Ta_i) = \sum_{n=1}^{3} Ta_i(\widehat{dv}_{mn}, \exists_n) \tag{7.12}$$

Step 8: Using Eq. (7.13), compute relative valuation $(\bar{\rho})$ score

$$(\bar{\rho}) = [\delta_{ik}]_{n \times n} \tag{7.13}$$

where $\delta_{ik} = (Eu_i - Eu_k) + \{\tau(Eu_i - Eu_k) \times (Ta_i - Ta_k)\}$, $k = 1,2\ldots n$.

$\tau \rightarrow$ threshold function and characterised as

$$\tau(j) = \begin{cases} 1 \text{ if } |j| \geq \vartheta \\ 0 \text{ if } |j| < \vartheta \end{cases} \tag{7.14}$$

$\vartheta \rightarrow$ threshold parameter and ranging from 0.01 to 0.05.

Step 9: Estimate the final valuation score (FVS_i) by using Eq. (15) and rank in the descending manner.

$$\text{FVS}_i = \sum_{k=1}^{n} \delta_{ik} \tag{7.15}$$

7.4   CASE STUDY

CPM, one of the important functionary systems of the considered industry, has been considered as a case for presenting the application of integrated FMEA and FCODAS approaches. As failure of this system has a direct effect on the delivery of milk and its products to the customers, therefore, for long run availability, risk analysis of CPM is highly essential. The application-based results are presented as follows:

7.4.1   Integrated FMEA – FCODAS Approach-Based Results

Under FMEA approach, selected team consisting of three experts was asked to execute a brainstorming session to list various failure causes connected to CPM, and an FMEA sheet was created. The three experts were requested to rate each specified failure cause under the three risk categories O_f, S, and O_d using linguistic scales given in Tables 7.1–7.3 and created FMEA sheet displayed in Table 7.4.

Under FCODAS approach, initial decision matrix was generated using Eq. (7.3) for specified failure causes under CPM utilising FCODAS application and three risk factor-related linguistic term scales (Tables 7.1–7.3). Table 7.6 shows the generated fuzzy decision matrix under CPM.

The average value of three risk factors was determined using Eq. (7.4) to account all experts' input, and the resulting average fuzzy decision matrix for CPM is shown in Table 7.7.

A crisp value decision matrix for the set of stated failure causes has been generated using average fuzzy values in Eqs. (7.5) and (7.6), as shown in Table 7.8.

Crisp decision matrix was utilised to tabulate normalised beneficial and non-beneficial decision matrices using Eq. (7.7) for specified failure causes under CPM. Non-beneficial criteria are O_f, and S, while beneficial criteria is O_d. Table 7.9 shows developed normalised beneficial and non-beneficial decision matrix under CPM.

TABLE 7.1

Failure Occurrence Scale Probability [30,31]

Linguistic Terms	Failure Occurrence (in months)	TFN
Very-2 high (VVH)	0–3	(8,9,10)
Very high (VH)	3–5	(6,7,8)
High (H)	5–8	(5,6,7)
Medium (M)	8–10	(4,5,6)
Fair (F)	10–12	(3,4,5)
Low (L)	12–24	(2,3,4)
Very low (VL)	>24	(1,2,3)

TABLE 7.2
Severity Scale [5,30]

Linguistic Terms	Severity Level	TFN
Extreme serious (ES)	Severity level is very high	(8,9,10)
Extreme (E)	High severity level with intimation	(7,8,9)
Major (MJ)	Due to component failure, the equipment is unavailable	(6,7,8)
Moderate (MD)	The performance of the system is harmed, and maintenance is required	(4,6,7)
Low (L)	Low maintenance necessity due to minor fault	(3,4,6)
No effect (N)	The performance of the system is unaffected	(1,2,3)

TABLE 7.3
Failure Detection Scale [30,31]

Linguistic Terms	Failure Detection Level	TFN
Very uncertain (VU)	Equipment disassembled and no visible failures were found, necessitating replacement	(7,8,10)
Uncertain (U)	Equipment disassembled and failure can be seen visually	(6,7,8)
Very remote (VR)	Automatic devices are used to identify failures	(4,5,6)
Remote (R)	Failure detection by display screen	(3,4,5)
Certain (C)	Failure detection very easy	(1,2,3)

Weighted normalised decision matrix $\widehat{dv}_{mn}$ has been developed for specified failure causes under CPM as per Eqs. (7.8) and (7.9). After consulting with maintenance professionals, the weights for tabulating weighted normalised matrix were determined to be 0.5, 0.3 and 0.2. Fuzzy negative-ideal solution ($\dot{\exists}$) values were computed from a weighted normalised decision matrix $\widehat{dv}_{mn}$ using Eq. (7.10). Furthermore, using $\dot{\exists}$ values in Eqs. (7.11) and (7.12), Euclidean (Eu_i) and Taxicab distance (Ta_i) were tabulated for specified failure causes under CPM shown in Table 7.10.

Using Eqs. (7.13)–(7.14), the relative valuation ($\bar{\rho}$) score for each specified failure causes was tabulated considering the threshold parameter (τ) constant value as 0.05. The relative valuation ($\bar{\rho}$) score for the set of specified failure causes is used in Eq. (7.15), and their final valuation scores (FVS_i) were tabulated. On the basis of tabulated scores, ranking of all specified failure causes was done in descending order as shown in Table 7.11.

TABLE 7.4
FMEA Sheet for CPM

S. No.	Components	Function	Failure Mode	Failure Effect	Failure Cause
1	Circular cutter	To cut the package horizontally	Mechanical failure	Hindrance in cutting process	Blunting of cutter (PKM1)
2	Cross cutter	To cut the package vertically	Mechanical failure	Hindrance in cutting process	Blunting of cutter (PKM2)
3	Brake	To stop operation	Friction fade	Related moving part will not stop	Wear (PKM3)
4	2/3 valve	To control the free movement of the block	Friction rupture	Uncontrolled movement of block	Wear (PKM4)
					Corrosion (PKM5)
5	Sensor	For controlling the movement of top roll through the identification mark	Open or short circuit	Uncontrolled movement of top roll	Failure of hardware (PKM6)
6	Shaft for circular cutter	To provide motion to circular cutter	Cracking/ bending of shaft	Breakdown	Overloading (PKM7)
					Overspeed (PKM8)
7	Shaft for cross cutter	To provide motion to cross cutter	Cracking/ bending of shaft	Breakdown	Overloading (PKM9)
					Overspeed (PKM10)
8	Motor for circular cutter	To give rotational motion to shaft of circular cutter	Mechanical/ electrical failure	No power transmission to shaft	Winding failure (PKM11)
9	Cross cutter motor	To give rotational motion to shaft of cross cutter	Mechanical/ electrical failure	No power transmission to shaft	Winding failure (PKM12)
10	Packaging roll motor	To give motion to packaging roll	Mechanical/ electrical failure	No power transmission to roll	Winding failure (PKM13)
11	Conveyor motor	To give motion to conveyor	Mechanical/ electrical failure	No power transmission to conveyor	Winding failure (PKM14)

TABLE 7.5
Expert's Feedback

S. No.	Failure Cause	O_f	S	O_d
1	PKM1	VH, H, VH	MJ, L, MD	U, C, VU
2	PKM2	VH, H, H	L, MJ, MD	U, C, U
3	PKM3	H, F, VH	MJ, MD, MJ	U, VU, VU
4	PKM4	H, M, F	L, MD, MD	C, U, C
5	PKM5	H, M, H	L, MD, L	C, C, U
6	PKM6	F, L, F	MJ, E, MJ	VU, U, U
7	PKM7	M, F, F	MJ, E, MJ	U, VU, U
8	PKM8	H, F, M	L, MD, MJ	C, U, C
9	PKM9	L, F, M	MJ, E, MJ	U, VU, U
10	PKM10	H, F, F	L, MD, MJ	C, U, C
11	PKM11	F, L, L	MJ, MD, MJ	VU, U, VU
12	PKM12	L, F, M	MJ, MJ, MD	VU, U, VU
13	PKM13	F, L, M	MJ, MD, MD	VU, U, VU
14	PKM14	L, F, L	MD, MJ, MD	VU, U, VU

TABLE 7.6
Initial Fuzzy Decision Matrix ($\widetilde{DV}$) under CPM

Failure Cause	O_f	S	O_d
PKM1	(6,7,8), (5,6,7), (6,7,8)	(6,7,8), (3,4,6), (4,6,7)	(6,7,8), (1,2,3), (7,8,10)
PKM2	(6,7,8), (5,6,7), (5,6,7)	(3,4,6), (6,7,8), (4,6,7)	(6,7,8), (1,2,3), (6,7,8)
PKM3	(5,6,7), (3,4,5), (6,7,8)	(6,7,8), (4,6,7), (6,7,8)	(6,7,8), (7,8,10), (7,8,10)
PKM4	(5,6,7), (4,5,6), (3,4,5)	(3,4,6), (4,6,7), (4,6,7)	(1,2,3), (6,7,8), (1,2,3)
PKM5	(5,6,7), (4,5,6), (5,6,7)	(3,4,6), (4,6,7), (3,4,6)	(1,2,3), (1,2,3), (6,7,8)
PKM6	(3,4,5), (2,3,4), (3,4,5)	(6,7,8), (7,8,9), (6,7,8)	(7,8,10), (6,7,8), (6,7,8)
PKM7	(4,5,6), (3,4,5), (3,4,5)	(6,7,8), (7,8,9), (6,7,8)	(6,7,8), (7,8,10), (6,7,8)
PKM8	(5,6,7), (3,4,5), (4,5,6)	(3,4,6), (4,6,7), (6,7,8)	(1,2,3), (6,7,8), (1,2,3)
PKM9	(2,3,4), (3,4,5), (4,5,6)	(6,7,8), (7,8,9), (6,7,8)	(6,7,8), (7,8,10), (6,7,8)
PKM10	(5,6,7), (3,4,5), (3,4,5)	(3,4,6), (4,6,7), (6,7,8)	(1,2,3), (6,7,8), (1,2,3)
PKM11	(3,4,5), (2,3,4), (2,3,4)	(6,7,8), (4,6,7), (6,7,8)	(7,8,10), (6,7,8), (7,8,10)
PKM12	(2,3,4), (3,4,5), (4,5,6)	(6,7,8), (6,7,8), (4,6,7)	(7,8,10), (6,7,8), (7,8,10)
PKM13	(3,4,5), (2,3,4), (4,5,6)	(6,7,8), (4,6,7), (4,6,7)	(7,8,10), (6,7,8), (7,8,10)
PKM14	(2,3,4), (3,4,5), (2,3,4)	(4,6,7), (6,7,8), (4,6,7)	(7,8,10), (6,7,8), (7,8,10)

TABLE 7.7
Average Fuzzy Decision Matrix

Failure Cause	O_f	S	O_d
PKM1	(5.667, 6.667, 7.667)	(4.333, 5.667, 7)	(4.667, 5.667, 7)
PKM2	(5.333, 6.333, 7.333)	(4.333, 5.667, 7)	(4.333, 5.333, 6.333)
PKM3	(4.667, 5.667, 6.667)	(5.333, 6.667, 7.667)	(6.667, 7.667, 9.333)
PKM4	(4,5,6)	(3.667, 5.333, 6.667)	(2.667, 3.667, 4.667)
PKM5	(4.667, 5.667, 6.667)	(3.333, 4.667, 6.333)	(2.667, 3.667, 4.667)
PKM6	(2.667, 3.667, 4.667)	(6.333, 7.333, 8.333)	(6.333, 7.333, 8.667)
PKM7	(3.333, 4.333, 5.333)	(6.333, 7.333, 8.333)	(6.333, 7.333, 8.667)
PKM8	(4,5,6)	(4.333, 5.667, 7)	(2.667, 3.667, 4.667)
PKM9	(3,4,5)	(6.333, 7.333, 8.333)	(6.333, 7.333, 8.667)
PKM10	(3.667, 4.667, 5.667)	(4.333, 5.667, 7)	(2.667, 3.667, 4.667)
PKM11	(2.333, 3.333, 4.333)	(5.333, 6.667, 7.667)	(6.667, 7.667, 9.333)
PKM12	(3,4,5)	(5.333, 6.667, 7.667)	(6.667, 7.667, 9.333)
PKM13	(3,4,5)	(4.667, 6.333, 7.333)	(6.667, 7.667, 9.333)
PKM14	(2.333, 3.333, 4.333)	(4.667, 6.333, 7.333)	(6.667, 7.667, 9.333)

TABLE 7.8
Crisp Value Decision Matrix

Failure Cause	O_f	S	O_d
PKM1	6.667	5.667	5.722
PKM2	6.333	5.667	5.333
PKM3	5.667	6.611	7.778
PKM4	5.000	5.278	3.667
PKM5	5.667	4.722	3.667
PKM6	3.667	7.333	7.389
PKM7	4.333	7.333	7.389
PKM8	5.000	5.667	3.667
PKM9	4.000	7.333	7.389
PKM10	4.667	5.667	3.667
PKM11	3.333	6.611	7.778
PKM12	4.000	6.611	7.778
PKM13	4.000	6.222	7.778
PKM14	3.333	6.222	7.778

7.5   RESULT AND DISCUSSION

Table 7.11 shows that the failure cause PKM14 under CPM has been ranked as the most critical with rank 1 utilising the FCODAS approach with a final valuation score (FVS_i) of 1.603. As a result, more attentiveness is essential for this cause in order to keep the equipment running continually. Failure cause PKM2, on the other hand, with final valuation score (FVS_i) of −1.220, has been placed 14th, implying that it

TABLE 7.9
Normalised Beneficial and Non-Beneficial Decision Matrix

Failure Cause	O_f (Non-Beneficial)	S (Non-Beneficial)	O_d (Beneficial)
PKM1	0.5	0.833333	0.735714
PKM2	0.526316	0.833333	0.685714
PKM3	0.588235	0.714286	1
PKM4	0.666667	0.894737	0.471429
PKM5	0.588235	1	0.471429
PKM6	0.909091	0.643939	0.95
PKM7	0.769231	0.643939	0.95
PKM8	0.666667	0.833333	0.471429
PKM9	0.833333	0.643939	0.95
PKM10	0.714286	0.833333	0.471429
PKM11	1	0.714286	1
PKM12	0.833333	0.714286	1
PKM13	0.833333	0.758929	1
PKM14	1	0.758929	1

TABLE 7.10
Weighted Normalised Matrix, $\exists$, Eu_i and Ta_i for CPM

Failure Cause	O_f	S	O_d	Eu_i	Ta_i
PKM1	0.250	0.250	0.147	0.078	0.110
PKM2	0.263	0.250	0.137	0.072	0.113
PKM3	0.294	0.214	0.200	0.116	0.171
PKM4	0.333	0.268	0.094	0.112	0.159
PKM5	0.294	0.300	0.094	0.116	0.151
PKM6	0.455	0.193	0.190	0.226	0.300
PKM7	0.385	0.193	0.190	0.165	0.230
PKM8	0.333	0.250	0.094	0.101	0.140
PKM9	0.417	0.193	0.190	0.192	0.262
PKM10	0.357	0.250	0.094	0.121	0.164
PKM11	0.500	0.214	0.200	0.272	0.377
PKM12	0.417	0.214	0.200	0.198	0.293
PKM13	0.417	0.228	0.200	0.200	0.307
PKM14	0.500	0.228	0.200	0.274	0.390
$\exists$	**0.250**	**0.193**	**0.094**		

requires less attention in order to avert machine failure. Other failure causes such as PKM1, PKM3, PKM4, PKM5, PKM6, PKM7, PKM8, PKM9, PKM10, PKM11, PKM12 and PKM13 with their corresponding scores −1.147, −0.608, −0.666, −0.620, 0.925, 0.072, −0.825, 0.451, −0.541, 1.583, 0.540 and 0.567 were ranked 13th, 9th, 11th, 10th, 3rd, 7th, 12th, 6th, 8th, 2nd, 5th and 4th, respectively.

TABLE 7.11

Final Valuation Score (FVS_i) and Ranking of Failure Cause

Failure Cause	1	2	...	14	(FVS_i)	Rank
PKM1	0.000	0.005	...	−0.193	−1.147	13
PKM2	−0.005	0.000	...	−0.198	−1.220	14
PKM3	0.039	0.044	...	−0.155	−0.608	9
PKM4	0.035	0.040	...	−0.159	−0.666	11
PKM5	0.038	0.043	...	−0.156	−0.620	10
PKM6	0.150	0.155	...	−0.048	0.925	3
PKM7	0.088	0.093	...	−0.108	0.072	7
PKM8	0.023	0.029	...	−0.171	−0.825	12
PKM9	0.115	0.121	...	−0.081	0.451	6
PKM10	0.044	0.049	...	−0.151	−0.541	8
PKM11	0.197	0.203	...	−0.001	1.583	2
PKM12	0.122	0.127	...	−0.075	0.540	5
PKM13	0.124	0.129	...	−0.073	0.567	4
PKM14	0.199	0.204	...	0.000	1.603	1

7.6 CONCLUSION

The risk analysis of the CPM was represented by applying fuzzy FMEA approach integrated with FCODAS. With the usage of these approaches, it has been identified that failure cause winding failure (PKM14) of the conveyor motor being prioritised as the most critical. The current approach can be extended by considering the weight of risk factors, which is not taken into account in this study. This study can also be applied to other subsystems of the sugar mill sector, paper mill and all process industries in order to undertake risk assessments in a fuzzy environment. Furthermore, the ranking results might be compared to other fuzzy-based MCDM techniques.

REFERENCES

1. Sharma RK, Kumar D, Kumar P. Modeling and analysing system failure behaviour using RCA, FMEA and NHPPP models. *International Journal of Quality & Reliability Management*. 2007.
2. Kushwaha DK, Panchal D, Sachdeva A. Reliability analysis of cutting system of sugar industry using intuitionistic fuzzy Lambda–Tau approach. In The Handbook of Reliability, Maintenance, and System Safety through Mathematical Modeling, 2021 (pp. 65–77). Academic Press.
3. Panchal D, Kumar D. Risk analysis of compressor house unit in thermal power plant using integrated fuzzy FMEA and GRA approach. *International Journal of Industrial and Systems Engineering*. 2017;25(2):228–50.
4. Panchal D, Chatterjee P, Pamucar D, Yazdani M. A novel fuzzy-based structured framework for sustainable operation and environmental friendly production in coal-fired power industry. *International Journal of Intelligent Systems*. 2022;37(4):2706–38.

5. Gopal N, Panchal D, Tyagi M. RAM Analysis of industrial system of a chemical industry. In *Reliability and Risk Modeling of Engineering Systems*, 2021 (pp. 11–26). Springer, Cham.
6. Sharma RK, Sharma P. Integrated framework to optimize RAM and cost decisions in a process plant. *Journal of Loss Prevention in the Process Industries.* 2012;25(6):883–904.
7. Panchal D, Kumar D. Integrated framework for behaviour analysis in a process plant. *Journal of Loss Prevention in the Process Industries.* 2016;40:147–61.
8. Panchal D, Mangla SK, Tyagi M, Ram M. Risk analysis for clean and sustainable production in a urea fertilizer industry. *International Journal of Quality & Reliability Management.* 2018.
9. Kushwaha DK, Panchal D, Sachdeva A. Risk analysis of cutting system under intuitionistic fuzzy environment. *Reports in Mechanical Engineering.* 2020;1(1):162–73.
10. Chatterjee P, Chakraborty S. Development of a meta-model for the determination of technological value of cotton fiber using design of experiments and the TOPSIS method. *Journal of Natural Fibers.* 2018;15(6):882–95.
11. Agarwal S, Tyagi M, Garg RK. Commencement of green supply chain management barriers: A case of rubber industry. In *Advances in Manufacturing and Industrial Engineering*, 2021 (pp. 685–699). Springer, Singapore.
12. Kumar N, Tyagi M, Garg RK, Sachdeva A, Panchal D, A framework development and assessment for cold supply chain performance system: A case of vaccines. *Operations Management and Systems Engineering.* 2021:339–353.
13. Panchal D, Chatterjee P, Sharma R, Garg RK. Sustainable oil selection for cleaner production in Indian foundry industries: A three phase integrated decision-making framework. *Journal of Cleaner Production.* 2021;313:127827.
14. Kumar N, Tyagi M, Sachdeva A. Depiction of possible solutions to improve the cold supply chain performance system. *Journal of Advances in Management Research.* 2021.
15. Petrović G, Mihajlović J, Ćojbašić Ž, Madić M, Marinković D. Comparison of three fuzzy MCDM methods for solving the supplier selection problem. *Facta Universitatis, Series: Mechanical Engineering.* 2019;17(3):455–69.
16. Zadeh LA. Fuzzy sets. *Information and Control.* 1965;8(3), 338–353.
17. Zimmermann, H., 1996. *Fuzzy Set Theory and Its Applications.* Kluwer Academic Publishers, London.
18. Đalić I, Ateljević J, Stević Ž, Terzić S. An integrated swot–fuzzy piprecia model for analysis of competitiveness in order to improve logistics performances. *Facta Universitatis, Series: Mechanical Engineering.* 2020;18(3) 439–51.
19. Ramakrishnan KR, Chakraborty S. A cloud TOPSIS model for green supplier selection. *Facta Universitatis, Series: Mechanical Engineering.* 2020;18(3):375–97.
20. Wang YM, Chin KS, Poon GK, Yang JB. Risk evaluation in failure mode and effects analysis using fuzzy weighted geometric mean. *Expert Systems with Applications.* 2009;36(2):1195–207.
21. Zhou Q, Thai VV. Fuzzy and grey theories in failure mode and effect analysis for tanker equipment failure prediction. *Safety Science.* 2016;83:74–9.
22. Mohsen O, Fereshteh N. An extended VIKOR method based on entropy measure for the failure modes risk assessment–A case study of the geothermal power plant (GPP). *Safety Science.* 2017;92:160–72.
23. Karatop B, Taşkan B, Adar E, Kubat C. Decision analysis related to the renewable energy investments in Turkey based on a Fuzzy AHP-EDAS-Fuzzy FMEA approach. *Computers & Industrial Engineering.* 2021;151:106958.
24. Das I, Panchal D, Tyagi M. A novel PFMEA-Doubly TOPSIS approach-based decision support system for risk analysis in milk process industry. *International Journal of Quality & Reliability Management.* 2021.

25. Liu HC, Chen XQ, Duan CY, Wang YM. Failure mode and effect analysis using multi-criteria decision making methods: A systematic literature review. *Computers & Industrial Engineering.* 2019;135:881–97.
26. Gopal N, Panchal D. A structured framework for performance optimization using JBLTO, FCOPRAS and FCODAS methodologies. *Journal of Quality in Maintenance Engineering.* 2022.
27. Keshavarz Ghorabaee M, Zavadskas EK, Turskis Z, Antucheviciene J. A new combinative distance-based assessment (CODAS) method for multi-criteria decision-making. *Economic Computation & Economic Cybernetics Studies & Research.* 2016;50(3).
28. Karaşan A, Boltürk E, Kahraman C. A novel neutrosophic CODAS method: Selection among wind energy plant locations. *Journal of Intelligent & Fuzzy Systems.* 2019;36(2):1491–504.
29. Maghsoodi AI, Rasoulipanah H, López LM, Liao H, Zavadskas EK. Integrating interval-valued multi-granular 2-tuple linguistic BWM-CODAS approach with target-based attributes: Site selection for a construction project. *Computers & Industrial Engineering.* 2020;139:106147.
30. Gopal N, Panchal D. A structured framework for reliability and risk evaluation in the milk process industry under fuzzy environment. *Facta Universitatis, Series: Mechanical Engineering.* 2021;19(2): 307–33.
31. Panchal D, Singh AK, Chatterjee P, Zavadskas EK, Keshavarz-Ghorabaee M. A new fuzzy methodology-based structured framework for RAM and risk analysis. *Applied Soft Computing.* 2019;74:242–54.

8 Performance Modelling and Performability Analysis of Repairable Industrial Systems Using Stochastic Petri Nets
An Overview

Sudhir Kumar and P.C. Tewari
National Institute of Technology

CONTENTS

8.1 INTRODUCTION

The tremendous evolution in the complexity in industrial systems is a challenging task for the engineers for analysing and controlling these complex systems. Direct analysis of the behaviour is not possible due to complexity and safety of both operators and systems. As the modern industrial systems are becoming more complex in nature, it is thus necessary to select the optimal design and maintenance policy for which proper modelling and analysis of the system are the basic requirements. Keeping in view these constraints, modelling and analysis of these systems using some tools and techniques such as Fault Tree Analysis (FTA), Reliability Block Diagram (RBD), Markovian Approach and Petri nets are available [1], having their specific advantages and limitations. The focus must be put on correctness of models as it ultimately affects the operational efficiency and thus the cost and time. Petri nets have the capability for modelling of such systems having concurrency, asynchronisation, parallelism random failure and stochastic behaviour [2,3]. As Petri nets have generality and permissiveness inherent, thus it can be used for a very wide variety of applications [4]. The technique has its usefulness for the systems which can be represented graphically like flow charts and need to represent parallel or concurrent activities [5].

8.2   DESCRIPTION OF PETRI NETS

A Petri net modelling essentially comprises four components, namely, places, transitions, arcs and tokens [6]. The directed arcs connect places to transitions and vice versa. In Petri nets, circles represent the places, bars represent the transitions and the transitions represent the event. Places represent the pre- and post-condition of the system [7].

- Places generally represent the conditions of the systems being modelled.
- Firing of transitions means the occurring of events in the system, i.e. modulation in the condition of the system.
- Arcs connect places to transitions and vice versa, connections between place to place and transition to transition are not allowed in the Petri nets. In simple, we can draw the Petri nets diagram with the help of transitions with its input and output places connecting them by the incoming and outgoing arcs.

The places from where transition receives the arc are called input places, and the places to which transition delivers the arc are called the output places of the transition.

Figure 8.1 illustrates the various elements required for the Petri Nets modelling of the subsystem.

A Petri net has five main elements, $PN = (P, T, I, O, M_O)$ where:

$P = \{P1\ P2\ P3\ P4\ P5...Pn\}$ is a non-empty finite set of places,

$T = \{T1\ T2\ T3\ T4\ T5...Tn\}$ is a non-empty finite set of transitions (there is no common element in the set P and T),

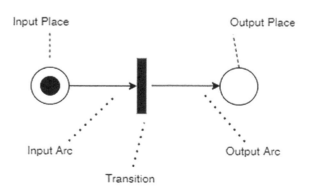

FIGURE 8.1   Petri net Diagram

I: $(P \times T) \rightarrow N$ is called the input function, where *N* is a set of non-negative integers and value *I* represents the number of directed arcs from places *p* to transitions *t*,

O: $(P \times T) \rightarrow N$ is called the output function, where *N* is a set of $=f$ non-negative integers, and value *O* represents the number of directed arcs from transitions *t* to places *p*,

M_O: $P \rightarrow N$ represents the initial marking of the model (Figure 8.2).

Input arcs: These are the directed arcs connecting the input place to transition, where it represents the guard function or the conditions need to be satisfied for the activation of transition (Figure 8.3a).

Output arcs: These are directed arcs connecting transitions to the output places, where it represents the post conditions resulting after the firing of transition (Figure 8.3b).

Input places: These are a set of places connected to transitions with input arcs (Figure 8.3c).

Output places: These are a set of places connected from the transitions with the help of output arcs (Figure 8.3d).

Tokens: Tokens are represented by black dots or integers carried by places; the presence of tokens in a place indicates the condition which the system holds (Figure 8.3e (left) and f (right)).

Marking: It indicates that the number and distribution of tokens over the various places will represent a configuration of the net called a marking.

$(m_1 m_2 \dots m_p)$; $P=$no. of places

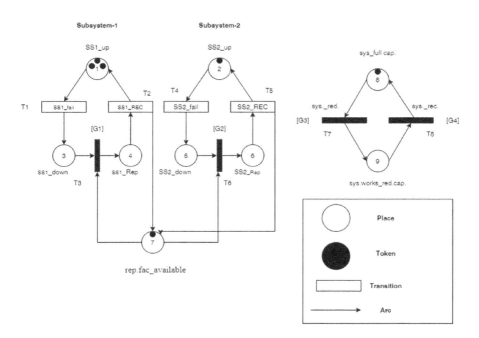

FIGURE 8.2 Example of Petri net having nine places and eight transitions having two subsystems.

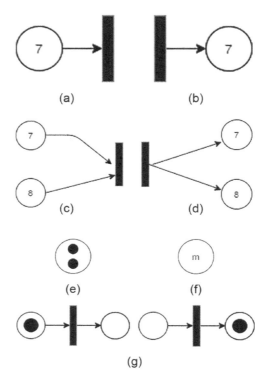

FIGURE 8.3   Various elements of Petri nets

The transition will only be fired if its input place carries the required number of tokens and satisfies the defined conditions (Figure 8.3g).

As the transition fires on happening of any event, it removes one token from all input places and deposits one token to all the output places it is connected (Figure 8.3h).

Guard functions: These are a defined set of rules imposed on transition for its firing.

8.3   EXAMPLE PERFORMANCE MODELLING OF REPAIRABLE INDUSTRIAL SYSTEM USING PETRI NETS

Consider an industrial system composed of five subsystems such as Condenser, Low-Pressure Heater (LPH), Condensate Extraction Pump (CEP), Boiler Feed Pump (BEP) and Deaerator. These subsystems are connected in series configurations. LPH, CEP and BFP have three parallel units each. The whole system can be modelled using the Petri Nets Tool [8,9]. Figure 8.4 illustrates the connection between various subsystems of the thermal power plant.

Figure 8.5 illustrates the Petri Nets Model of an industrial system (feed water system of the thermal power plant). The various places, transitions and guard functions in this model are as follows:

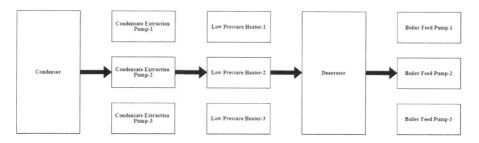

FIGURE 8.4 An industrial system (feed water system of thermal power plant).

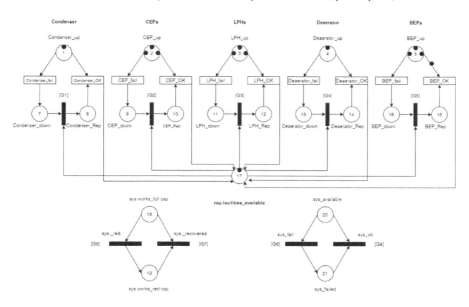

FIGURE 8.5 Petri nets model of an industrial system.

Places:

Condenser_up, CEP_up, LPH_up, Deaerator_up, BEP_up: These are the various places that represent the working state, i.e. upstate of condenser, CEPs, LPHs, Deaerator and BEPs.

Condenser_down, CEP_down, LPH_down, Deaerator_down, BEP_down: These are the places that represent the subsystems in downstate, i.e. condenser, CEPs, LPHs, Deaerator and BEPs are waiting for repair.

Condenser_Rep, CEP_Rep, LPH_Rep, Deaerator_Rep, BEP_Rep: These are the places that represent the subsystems under repair, i.e. condenser, CEPs, LPHs, Deaerator and BEPs are under repair.

sys_available: This represents a place where the whole system is available in the working state.

sys.works_full cap.: This represents the place where the whole system is working in full capacity.

sys.works_red.cap.: It is the place that represents the working of system in reduced capacity.

sys_failed: It is the place that represents the state of system where it is in a state of complete failure.

rep.facilities_available: This place represents the availability of immediate repair facility.

Transitions:
condenser_fail, CEP_fail, LPH_fail, Deaerator_fail, BEP_fail: These represent timid transition fires corresponding to failure or repair, distribution delay associated with failure pattern of condenser, CEPs, LPHs, Deaerator and BEPs.

condenser_OK, CEP_OK, LPH_OK, Deaerator_OK, BEP_OK: These represents timid transitions associated with failure pattern of condenser, CEPs, LPHs, Deaerator and BEPs.

rep.avail_Condenser, rep.avail_CEP, rep.avail_LPH rep.avail_Deaerator, rep.avail_BEP These are the transitions called immediate transitions, which instantaneously fire if the conditions of guard functions are met associated with the availability of repair facility for condenser, CEPs, LPHs, Deaerator and BEPs.

sys_red, sys_recovered, sys_fail, and **sys_ok:** These are also the immediate transitions, which fire immediately without any time delay when it works in full and reduced capacity.

Assumptions while Modelling of Industrial System:
The assumptions made during the performance modelling of system using Petri nets [10,11] are as follows:

- The behaviour of Standby systems is similar in nature to that of active system;
- Before the failure of the whole system, it may work in reduced capacity;
- Simultaneous failures in the subsystems do not take place;
- Subsystem will restore its original functionality as a new one after the repair;
- Repair facilities are available with any delay except the availability of repairman;
- Failure rate of subsystems follows the exponential distribution;
- Repair rate of subsystems follows the Weibull distribution;
- Service of subsystem means repair and replacements of parts;
- Failure/repair rates are constant over time and statistically independent;
- Availability of repair facilities includes repairman and resources required.

Initially, the system is in fully working state having five subsystems shown by five circles as Condenser_up, CEP_up, LPH_up, Deaerator_up and BEP_up as shown in Figure 8.5. These places carry one, three, three, one and three tokens (represented by black dots), respectively. The presence of one or more tokens at these places represents that the subsystems are in working state either in full capacity or reduced capacity. Each token represents the number of parallel units each subsystem carries

[12–14]. Tokens available at the place [15] that represents the availability of repair facilities available at a given point of time. Each subsystem undergoes failure with firing of transitions condenser_fail, CEP_fail, LPH_fail, Deaerator_fail and BEP_fail, respectively. These are the timid transitions having particular failure rate ($\mu_1, \mu_2, \mu_3, \mu_4, \mu_5$) as per the failure pattern of each subsystem [15–18]. On successive firing of the mentioned transitions, one token moves from an incoming place to an outgoing place of these transitions every time.

Distribution of the tokens at places changes with random failure of the components. After failure of subsystems, the token moves to places Condenser_down, CEP_down, LPH_down, Deaerator_down and BEP_down, with the presence of token at these places represents that subsystems are waiting for repair facilities as illustrated in Figure 8.6, and it will initiate the firing of transitions rep.avail_Condenser, rep.avail._CEP, rep.avail._LPH rep.avail_Deaerator and rep.avail_BEP, respectively. These are the immediate transitions, which will fire without delay after satisfying the guard functions and availability of the repair facilities with availability token at repair places [19,20]. In case of multiple failures of subsystems, one subsystem has to wait for the availability of repair facility.

Firing of transitions rep.avail_Condenser, rep.avail_CEP, rep.avail_LPH rep.avail_Deaerator and rep.avail_BEP will move the token at places Condenser_Rep, CEP_Rep, LPH_Rep, Deaerator_Rep and BEP_Rep, respectively. System is said to be under repair at these places as shown in Figure 8.7. Availability of tokens at these places will result in the firing of timid transitions condenser_OK, CEP_OK, LPH_OK, Deaerator_OK and BEP_OK, respectively. These are also the timid transitions that will fire on the availability of token at the incoming places with some repair rates

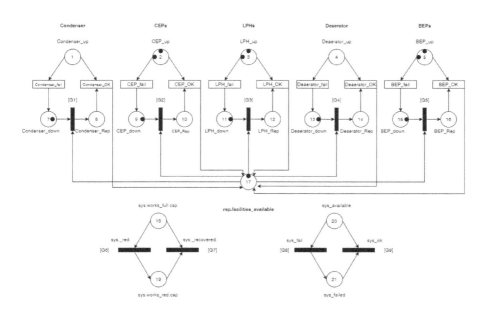

FIGURE 8.6 Downstate of feed water system (waiting for repair facilities).

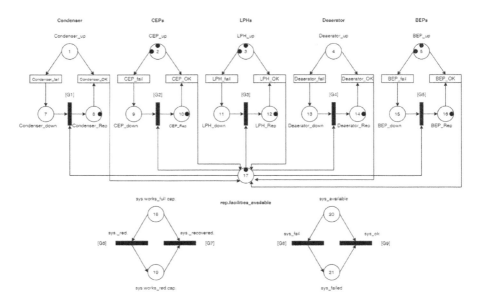

FIGURE 8.7 Subsystems after firing of immediate transitions (under repair).

$(\rho_1, \rho_2, \rho_3, \rho_4, \rho_5)$ based upon the repair schedule and time. Firing of these transitions will shift the system into working state as the initial marking.

Further, as we have seen that some places, i.e., 2,3,5, carry multiple tokens means these have parallel units, and failure of any one of these subsystems will bring the whole system into reduced capacity (not complete failure), whereas the failure of subsystems 1 and 4 will bring the subsystem into a complete failure state.

Each transition is assigned with some rules of firing called guard function [21]. The various guard functions assigned to various transitions are as follows:

Programmed Guard Functions:
All the transitions are provided with a set rule of firing called the guard functions. The guard functions for different transitions are as follows:

[G 1]: = (#7>0 and #17>0) enables the firing of transition rep. avail_Condenser.
[G 2]: = (#9>0 and #17>0) enables the firing of transition rep. avail_CEP.
[G 3]: = (#11>0 and #17>0) enables the firing of transition rep. avail_LPH.
[G 4]: = (#13>0 and #17>0) enables the firing of transition rep. avail_Deaerator.
[G 5]: = (#15>0 and #17>0) enables the firing of transition rep. avail_BEP.
[G 6]: = #2<3 and #2>0 or #3<3 and #3>0 or #5<3 and #5>0) enables the firing of transition sys_red.
[G 7]: =(#2>2and#3>2and#5>2) disables the firing of transition sys_recovered.
[G 8]: = (#1>0 or #2>0, or #2>3, or #4>0, or #5>0) enables the firing of transition sys_fail.
[G 9]: = (#1>0 and #2>0, and #2>3, and #4>0, and #5>0) disables the firing of transition sys_ok.

8.4   FAILURE MODELLING

After the successful modelling of the system concerned with use of Petri nets, it is required to make the performance analysis of the same.

The present approach of using Petri nets for modelling of the system permits to model the failure through the probability density function of a stochastic distribution for each subsystem [22–24]. For the specific test case, a probability density function of an exponential distribution can be used as

$$F(t) = \mu e^{-\mu t}$$

where μ is the failure rate computed as

$$\mu = \frac{1}{\text{MTBF}}$$

The failure rates can be derived from MTBF values, and it is possible to observe the failure distribution in a specific time interval for a Monte Carlo simulation.

8.5   PERFORMANCE ANALYSIS

Performance analysis was carried out by a measure of the performability of the system concerned as a function of availability with the use of Petri nets approach. Performability was evaluated to observe the long-run performance as a function of availability [25–27]. For this purpose, licensed software GRIF-predicates were used. Failure and repair rates (FRRs) are assumed to follow the Weibull and exponential patterns, respectively, in the present study. The behavioural characteristics of plant were determined with Monte Carlo simulation by running the same up to 10,000 hours and 21,000 replications with a confidence level of 95% [28]. Performability matrices are obtained with combinations of various FRRs of each subsystem keeping parameters constant. MATLAB software was used for solving the performability levels of each subsystem [29]. In order to analyse the performance, performability matrices and the figures for various subsystems of the feed water system, which shows the impact of numerous FRRs on the performability of the concerned system, are given as follows.

Table 8.1 and Figure 8.8 show the impact of varying FRRs of condenser on the performability of system concerned as a function of availability. It is clearly seen that for some known value of failure (μ) and repair rate (ρ), there is a sharp decrease in the availability of the system with increase in the failure rate and vice versa. Approximately, 20% change is observed in the overall availability of the system with a variation in combinations of FRR.

Table 8.2 and Figure 8.9 reveal the variation in availability of CEPs with combinations of FRR. It is observed that there is a sharp variation in the availability with change in FRRs. An overall 10% variation is observed in the availability of the system concerned.

Table 8.3 and Figure 8.10 reveal the pattern of variation in the availability with different FRRs of LPHs. A slight change in the availability is observed for different

TABLE 8.1

Performability-Matrix for Condenser Subsystem

$\mu 1$ \ $\rho 1$	0.055	0.060	0.065	0.070	0.075	Constant Parameters
0.005	0.8887	0.8933	0.8965	0.9071	0.9100	$\mu_2=0.014\ \rho_2=0.15$
0.010	0.8165	0.8233	0.8423	0.8447	0.8451	$\mu_3=0.004\ \rho_3=0.15$
0.015	0.7503	0.7598	0.7732	0.7840	0.7982	$\mu_4=0.0025\ \rho_4=0.125$
0.020	0.6950	0.6956	0.7296	0.7415	0.7492	$\mu_5=0.00005\ \rho_5=0.015$
0.025	0.6483	0.6639	0.682	0.6906	0.7101	

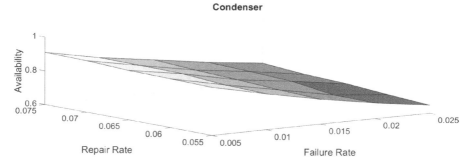

FIGURE 8.8   Impact of changing FRR of condenser on the performability.

TABLE 8.2

Performability-Matrix for Condensate Extraction Pump

$\mu 2$ \ $\rho 2$	0.05	0.100	0.15	0.200	0.25	Constant Parameters
0.012	0.7167	0.7711	0.7816	0.7835	0.7884	$\mu_1=0.015\ \rho_1=0.065$
0.013	0.7111	0.7677	0.7725	0.779	0.7836	$\mu_3=0.004\ \rho_3=0.15$
0.014	0.7005	0.7562	0.7732	0.7859	0.7869	$\mu_4=0.0025\ \rho_4=0.125$
0.015	0.7027	0.7601	0.7771	0.7809	0.7818	$\mu_5=0.00005\ \rho_5=0.015$
0.016	0.6834	0.7516	0.774	0.7805	0.7814	

combinations of failure (μ) and repair rate (ρ), and it is clearly seen that as the failure rate (FR) increases from 0.002 to 0.026, the availability decreases sharply from 76.73% to 74.50% (2.25% approx.), whereas the overall availability of Feed Water System (FWS) varies from 74.50% to 77.76% with different combinations of FRRs of LPHs.

Similarly, Table 8.4 and Figure 8.11 illustrate the impact of variation in FRR of deaerator on the performability of the system in terms of availability. The availability of the system decreases with increase in the failure rate and vice versa, whereas

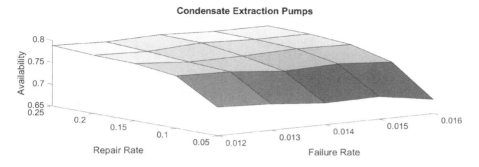

FIGURE 8.9 Impact of changing FRR of condensate extraction pump on the availability.

TABLE 8.3
Performability-Matrix for Low-Pressure Heater

$\mu3$ \ $\rho3$	0.05	0.1	0.15	0.2	0.25	Constant Parameters
0.002	0.7673	0.7689	0.7737	0.7745	0.7776	$\mu_1=0.015\ \rho_1=0.065$
0.003	0.7634	0.7681	0.7735	0.7739	0.7763	$\mu_2=0.014\ \rho_2=0.15$
0.004	0.7589	0.7711	0.7732	0.7733	0.7757	$\mu_4=0.0025\ \rho_4=0.125$
0.005	0.7569	0.7708	0.7718	0.7732	0.7743	$\mu_5=0.00005\ \rho_5=0.015$
0.006	0.745	0.7715	0.7717	0.7731	0.774	

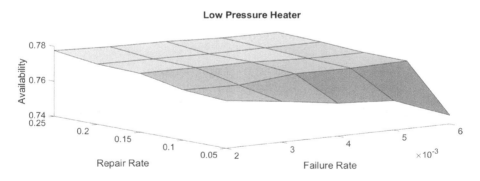

FIGURE 8.10 Impact of changing FRR of low-pressure heater on the availability.

the availability of the system increases with increase in the repair rate. A detectable change is observed in the overall availability with variation in FRR.

Table 8.5 and Figure 8.12 describe the impact of varying FRRs of BEP subsystem on the performance of system in terms of availability. The availability of the system increases with increase in the repair rate and decreases with increase in the failure rate. A detectable change in the overall availability is observed with variation of FRR.

TABLE 8.4
Performability-Matrix for Deaerator

$\mu 4$ \ $\rho 4$	0.115	0.12	0.125	0.13	0.135	Constant Parameters
0.0023	0.7734	0.7751	0.781	0.7813	0.785	$\mu_1 = 0.015\ \rho_1 = 0.065$
0.0024	0.7733	0.7745	0.7768	0.7779	0.7796	$\mu_2 = 0.014\ \rho_2 = 0.15$
0.0025	0.7731	0.7741	0.7732	0.7772	0.7773	$\mu_3 = 0.004\ \rho_3 = 0.15$
0.0026	0.7715	0.7716	0.7716	0.7753	0.7748	$\mu_5 = 0.00005\ \rho_5 = 0.015$
0.0027	0.7714	0.7715	0.7727	0.7746	0.7747	

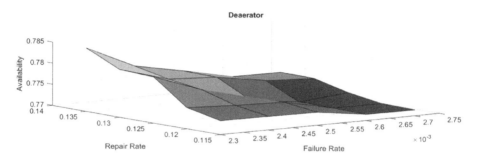

FIGURE 8.11   Impact of changing FRR of deaerator on the availability.

TABLE 8.5
Performability-Matrix for Boiler Feed Pumps

$\mu 5$ \ $\rho 5$	0.009	0.012	0.015	0.018	0.021	Constant Parameters
0.00003	0.7763	0.7764	0.7776	0.7779	0.7784	$\mu_1 = 0.015\ \rho_1 = 0.065$
0.00004	0.7756	0.7762	0.7762	0.7771	0.7773	$\mu_2 = 0.014\ \rho_2 = 0.15$
0.00005	0.773	0.7731	0.7732	0.7738	0.7757	$\mu_3 = 0.004\ \rho_3 = 0.15$
0.00006	0.7712	0.7731	0.7732	0.7733	0.7757	$\mu_4 = 0.0025\ \rho_4 = 0.125$
0.00007	0.7620	0.764	0.768	0.7715	0.7718	

The availability of the system is determined with variation in the availability of repair facilities present in the plant as seen in Table 8.6 and Figure 8.13. It is observed that availability of the system increases with increase in the repair facilities initially, but the variation and pattern of variation become constant after a particular value of repair facilities. It means that there is no need of further increase in the repair facilities as availability of the system remains constant for further increase. This will not be economical in terms of overall cost of maintenance and will increase without further increase in the overall availability of the system concerned.

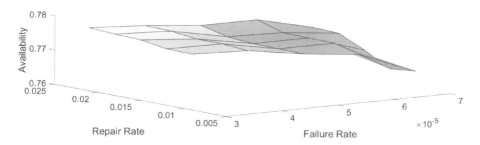

FIGURE 8.12 Impact of changing FRR of boiler feed pumps on the availability.

TABLE 8.6
Variation in the Overall Performability of Steam Generation System with Increase in Repair Facilities

No. of Repair Facilities	1	2	3	4	5
Availability	0.7732	0.7820	0.7890	0.7892	0.7892

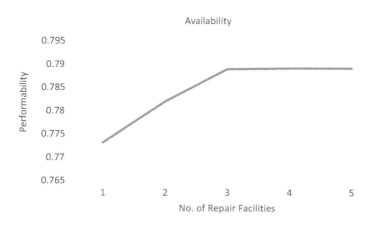

FIGURE 8.13 Variation in the availability with no. of repair facilities.

8.6 RESULTS AND DISCUSSIONS

The performability matrices as given by Tables 8.1–8.5 represent the behaviour of feed water system performability with the change in the FRR of different subsystems. These matrices help to propose the maintenance priorities to consider the criticality of various subsystems, which will further help the maintenance engineers to

TABLE 8.7
Priorities of Maintenance for Various Subsystems of Feed Water System

Name of Subsystem	Variation in Failure Rates (μ)	Variation in Repair Rates (ρ)	% Change in Availability	Repair Priority
Condenser	0.005–0.025	0.055–0.075	64.8%–91%	1
Low-pressure heater	0.002–0.026	0.05–0.25	74.50%–77.76%	3
Condensate extraction pump	0.012–0.016	0.05–0.25	68.34%–78.84%	2
Deaerator	0.0023–0.0027	0.115–0.135	77.14%–78.50%	5
Boiler feed pump	0.00003–0.00007	0.009–0.021	76.20%–77.84%	4

take the corrective action timely for the most critical system. As discussed above, the overall availability of FWS varies from 64.8% to 91% with different combinations of FRRs of condenser. Similarly, with varying FRRs of CEPs, the availability of FWS varies from 68.34% to 78.84%. For LPHs, it varies from 74.50% to 77.76%. There is a noticeable change in the availability of Deaerator and BEPs by varying the combination of FRRs. For Deaerator, it varies from 77.14% to 78.50%, and for BEPs, the variation is from 76.20% to 77.84%. The maintenance priorities for various subsystems of the FWS are mentioned in Table 8.7 as per the following order.

It is seen in Table 8.7 that the availability of condenser varies maximum with variation in FRR; hence, it is said to be the most critical subsystem, and in the same manner, availability of Deaerator varies very less, and hence it is said to be the least critical subsystem. So, while deciding the maintenance policies, the condenser must be kept at the top most priority and the Deaerator can be kept at the least priority.

REFERENCES

[1] K.S. Trivedi, V. Jindal, S. Dharmaraja, Stochastic modelling techniques for secure and survivable systems, In *The Morgan Kaufmann Series in Networking, Information Assurance*, Morgan Kaufmann, ISBN 9780123735669, pp. 171–207, 2008. USA https://doi.org/10.1016/B978-012373566-9.50009-4.

[2] Y. Narahari, N. Vishwanadham, A Petri net approach to the modelling and analysis of flexible manufacturing systems, *Annal of Operations Research*, 3, 449–472, 1985. https://doi.org/10.1007/BF02023780.

[3] N. Viswanadham, Y. Narahari, *Performance Modelling of Automated Manufacturing Systems*, PHI Learning Pvt. Ltd, Delhi 2015.

[4] T. Murata, Petri nets: Properties, analysis and applications, *Proceedings of the IEEE*, 77(4), 1989.

[5] J. Wang, Petri nets, *Handbook of Finite State Based Models and Applications*, Chapman and Hall/CRC Press, New York 2012.

[6] R. Zurawski, M. Zhou, Petri nets and industrial applications: A tutorial, *IEEE Transactions on Industrial Electronics*, 41(6), 1994.

[7] K. Jensen, Colored petri nets and the invariant method, *Theoretical Computer Science*, 14, 317–336, 1981. https://doi.org/10.1016/0304-3975(81)90049-9.

[8] J.S. Tan, M.A. Kramer, A general framework for preventive maintenance optimization in chemical process operations, *Computers and Chemical Engineering*, 21, 1451–1469, 1997. https://doi.org/10.1016/S0098-1354(97)88493-1.

[9] N. Gdownta, M. Saini, A. Kumar, Operational availability analysis of generators in steam turbine power plants, *SN Applied Sciences*, 2(779), 2020. https://doi.org/10.1007/s42452-020-2520-y.

[10] A. Sachdeva, D. Kumar, P. Kumar, Reliability analysis of the pulping system using Petri Nets, *International Journal of Quality & Reliability Management*, 25(8), 860–877, 2008. https://doi.org/10.1108/02656710810898667.

[11] A. Sachdeva, P. Kumar, D. Kumar, Behavioral and performance analysis of feeding system using stochastic reward nets, *The International Journal of Advanced Manufacturing Technology*, 45, 156–169, 2009. https://doi.org/10.1007/s00170-009-1960-8.

[12] G. Thangamani, Generalized stochastic petri nets for reliability analysis of lube oil system with common-cause failures, *American Journal of Computational & Applied Mathematics*, 2(4), 152–158, 2012. https://doi.org/10.5923/j.ajcam.20120204.03.

[13] H. Li, H. Zheng, H. Zhao, Z. Zheng, Research on the availability analysis method of navigation satellite based on petri nets, *China Satellite Navigation Conference (CSNC) 2018 Proceedings*, 127–136, 2018. https://doi.org/10.1007/978-981-13-0029-5_12.

[14] A.S. Angel, R. Jayaparvathy, Performance modelling of an intelligent emergency evacuation system in buildings on accidental fire occurrence, *Safety Science*, 12, 196–205, 2019. https://doi.org/10.1016/j.ssci.2018.10.027.

[15] J. Singh, S. Garg, Availability analysis of core veneer manufacturing system in plywood industry, *International Conference on Reliability and Safety Engineering*, Indian Institute of Technology, Kharagpur, 497–508, 2005.

[16] D. Kumar, I.P. Singh, J. Singh, Reliability analysis of the feeding system in the paper industry, *Microelectronics Reliability*, 28, 213–215, 1988. https://doi.org/10.1016/0026-2714(88)90353-8.

[17] B. Peros, Reliability analysis of the structures with a dominant wind load, *International Journal for Engineering Modelling*, 5(3–4), 1992.

[18] R. Dekker, W. Groenendijk, Availability assessment methods and their application in Practice, *Microelectronics Reliability*, 35(9–10), 1257–1274, 1995. https://doi.org/10.1016/0026-2714(95)99376-T.

[19] C.E. Okafor, A.A. Atikpakpa, U.C. Okonkwo, Availability assessment of steam and gas turbine units of a thermal power station using Markovian approach, *Archives of Current Research International*, 6(4), 1–17, 2016. https://doi.org/10.9734/ACRI/2016/30240.

[20] S. Gdownta, P.C. Tewari, Simulation modeling and analysis of complex system of thermal power plant, *Journal of Industrial Engineering and Management*, 2(2), 387–406, 2009. https://doi.org/10.3926/jiem.2009.v2n2.

[21] S.P. Sharma, H. Garg, Behavioural analysis of urea decomposition system in a fertiliser plant, *International Journal of Industrial and Systems Engineering*, 8(3), 2011.

[22] N. Kumar, P.C. Tewari, A. Sachdeva, Performance modelling and analysis of refrigeration system of a milk processing plant using petri nets, *International Journal of Performability Engineering*, 15(7), 1751–1759, 2019. https://doi.org/10.23940/ijpe.19.07.

[23] A.M. Dalavi, P.J. Pawar, T.P. Singh, Determination of optimal tool path in drilling operation using Modified Shuffled Frog Leaping Algorithm, *International Journal for Engineering Modelling*, 32(2–4), 33–44, 2019. https://doi.org/10.31534/eng-mod.2019.2-4.ri.01v.

[24] S. Malik, P.C. Tewari, Performance modelling and maintenance priorities decision for water flow system of a coal based thermal power plant, *International Journal of Quality and Reliability Management*, 35(4), 1–16, 2018. https://doi.org/10.1108/IJQRM-03-2017-0037.

[25] A. Kumar, V. Modgil, Performance optimization for ethanol manufacturing system of distillery plant using particle swarm optimization algorithm, *International Journal of Intelligent Enterprise*, 5(4), 345–364, 2018. https://doi.org/10.1504/IJIE.2018.095723.

[26] N. Tomasz, W.W. Sylwia, M. Chlebus, Reliability assessment of production process – Markov modelling approach, *Intelligent Systems in Production Engineering and Maintenance – ISPEM 2017. ISPEM 2017. Advances in Intelligent Systems and Computing*, Vol. 637. Springer, Cham, 2018.

[27] O. Dahiya, A. Kumar, M. Saini, Mathematical modeling and performance evaluation of apan crystallization system in a sugar industry, *SN Applied Sciences*, 2019. https://doi.org/10.1007/s42452-019-0348-0.

[28] M.A. Marsan, G. Conte, G. Balbo, A class of generalized stochastic petri nets for the performance evaluation of multiprocessor systems, *ACM transactions on Computer Systems*, 2(2), 93–122, 1984. https://doi.org/10.1145/190.191.

[29] J. Dugan, K. Trivedi, R. Geist, V. Nicola, *Extended Stochastic Petri Nets: Application and Analysis, Proceedings of the Performance '84, Paris, France*, 507–519, 1984. https://doi.org/10.21236/ADA148439.

9 Performance Analysis of Polyethylene Terephthalate Bottle Hot Drink Filling System by Fuzzy Availability

Pawan Kumar, Parul Punia, and Amit Raj
Central University of Haryana

CONTENTS

9.1 INTRODUCTION

Reliability is a measure of how likely it is that a product will function or service will be provided properly under the intended operating conditions for a fixed period. In other words, the ability of a function to remain operational during its design life is known as reliability [1]. Modern engineering systems are becoming more complex, posing greater risks in maintaining the production of the system, hence making reliability an important aspect when designing an overall system. Reliability of a

DOI: 10.1201/9781003140092-9

component is not specific, and it depends upon various environmental and operational conditions such as adverse operating conditions, age and unpredictable manufacturing processes affecting each component of the system differently [2]. There always lies an uncertainty in determining the future performance of the system as it is measured using failure and repair rates which are collected from the available past records of the system. To overcome such issues of uncertainty, fuzzy concept was proposed by L.A. Zadeh in 1965, suggesting the replacement of binary state assumptions with fuzzy state assumptions in conventional reliability [3,4]. The reliability concept based on both probabilistic and fuzzy state assumptions is known as profust reliability [5,6]. In any industrial system, there are various subsystems arranged in series, parallel or a combination of both. Depending upon the working of the subsystems, the system is either fully functional, working in reduced state or stops working completely. The theory of fuzziness works effectively for a system from a fully operating to a completely downstate. S.G. Chowdhury and K.B. Misra presented a method for determining the fuzzy reliability of a non-series parallel network [7]. Jau-Chaun Ke et al. presented a method to construct the membership function for analysis of fuzzy steady-state availability [8]. V. Modgil and P. Kumar analyzed the time-dependent availability of an industrial system [9]. A. Kumar and M. Saini analyzed the fuzzy availability of a marine power plant [10]. Using a mixed redundancy strategy and heterogeneous components, Z. Ouyang et al. presented an improved particle swarm optimization algorithm for reliability-redundancy allocation problems [11]. P.K. Chhoker and A. Nagar presented a mathematical model and analyzed the fuzzy availability of a stainless steel utensil manufacturing unit in steady state [12]. C. Kai-Yuan et al. presented a fuzzy reliability model of gracefully degradable computing systems [13].

In this chapter, we consider polyethylene terephthalate (PET) bottle hot drink filling system and solve the governing differential equation in MATLAB R2021a to calculate the fuzzy availability of the system in the transient state by varying failure and repair rate at a certain system coverage factor.

In this chapter, Section 9.1 is introductory. Section 9.2 provides a brief description of the system, various notations of the subsystem along with basic assumptions considered for the analysis. In Section 9.3, the mathematical formulation of the Chapman–Kolmogorov differential equation of the PET bottle hot drink filling system is done, where variable failure and repair rates are assumed. In Section 9.4, time-dependent system fuzzy availability analysis is done using various combinations of failure and repair rates. In Section 9.5, the conclusion drawn from the analysis is discussed.

9.2 SYSTEM DESCRIPTION, NOTATIONS AND ASSUMPTIONS

The PET bottle hot drink filling system is divided into many subsystems such as blow molding machine, filling machine, cooling machine, coding machine, labeling machine and pasteurizer cum storage tank. The main five subsystems are described below:

1. **Blow molding machine (P):** The blow molding operation is performed on the raw material to make the PET bottle. Two blow molding machines are used for this purpose, connected in parallel. The system fails when both the units stop operating.

2. **Filling machine (Q):** After the making of PET bottles, the bottles are first cleaned and then filled with a hot drink (i.e. at 70°C) in measured quantity by the filling machine. After that, the bottles are capped and sealed. Only one machine is used for filling, and the system fails once the unit stops operating.
3. **Cooling machine (R):** After the bottles are properly filled, capped and sealed, cool water is sprinkled on the bottles to bring the temperature of the filled bottle down to the required temperature. Two machines are used for this purpose, connected in parallel and are subjected to major as well as minor failures. So, the system works in a reduced state, and the system fails when both the units stop operating.
4. **Coding machine (S):** Once the desired temperature of the bottles is obtained, the coding machine then prints all the details like manufacturing date, expiry date, price of the bottle and batch number on the filled bottle. Only one machine is used for coding and the system fails once the unit stops operating.
5. **Labeling machine (T):** Labeling machine pastes the label on the bottle. Only one machine is used for labeling and the system fails once the unit stops operating.

The pasteurizer/storage tank stores and heats the fruit pulp to 70°C and then transfer it to the filling machine. This subsystem is not analyzed as this is subjected to minor failures only.

Notations:
P, Q, R, S and T represent the working states of the systems, namely, blow molding machine, filling machine, cooling machine, coding machine and labeling machine, respectively.

p, q, r, s and t represent failed states of the systems, namely, blow molding machine, filling machine, cooling machine, coding machine and labeling machine, respectively.

$\bar{P}$ and $\bar{R}$ represent the reduced state of subsystems P and R, respectively.

$\lambda_1, \lambda_2, \lambda_3, \lambda_4, \lambda_5, \lambda_1$ and λ_3 represent the failure rate of $P, Q, R, S, T, \bar{P}$ and $\bar{R}$, respectively.

$\mu_1, \mu_2, \mu_3, \mu_4$ and μ_5 represent the repair rate of P, Q, R, S and T, respectively.

c represents the system coverage factor.

$A(t)$ represents the availability of the system at time t.

Assumptions

1. Initially, all the subsystems are in working condition.
2. Failure and repair rates follow the exponential distribution and are independent of each other.
3. Subsystems P and R fail through reduced state only.
4. Repaired unit works as good as the new one.

Based upon the notations and assumptions mentioned above, the state transition diagram of the system is given in Figure 9.1.

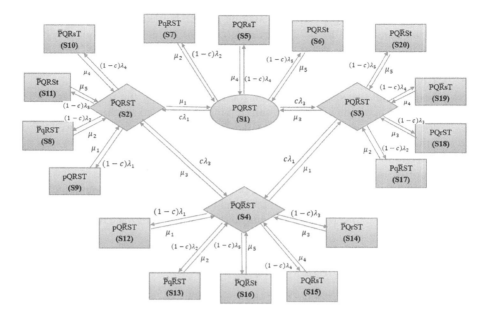

FIGURE 9.1   State transition diagram of the system.

9.3   MATHEMATICAL MODELING OF THE TRANSIENT STATE OF THE SYSTEM

Based on the mnemonic rule, mathematical modeling of the system is done. Using this mathematical modeling, differential equations associated with the transition diagram are developed.

The following are the first-order Chapman–Kolmogorov differential equations associated with the transition diagram:

$$P_1^{'}(t) = -KP_1(t) + \mu_1 P_2(t) + \mu_3 P_3(t) + \mu_2 P_7(t) + \mu_4 P_5(t) + \mu_5 P_6(t) \tag{9.1}$$

$$P_2^{'}(t) = -LP_2(t) + c\lambda_1 P_1(t) + \mu_2 P_8(t) + \mu_3 P_4(t)$$
$$+ \mu_4 P_{10}(t) + \mu_5 P_{11}(t) + \mu_1 P_9(t) \tag{9.2}$$

$$P_3^{'}(t) = -MP_3(t) + c\lambda_3 P_1(t) + \mu_2 P_{17}(t) + \mu_3 P_{18}(t)$$
$$+ \mu_4 P_{19}(t) + \mu_5 P_{20}(t) + \mu_1 P_4(t) \tag{9.3}$$

$$P_4^{'}(t) = -NP_4(t) + c\lambda_1 P_3(t) + c\lambda_3 P_2(t) + \mu_1 P_{12}(t)$$
$$+ \mu_2 P_{13}(t) + \mu_3 P_{14}(t) + \mu_4 P_{15}(t) + \mu_5 P_{16}(t) \tag{9.4}$$

$$P_5'(t) = -\mu_4 P_5(t) + (1-c)\lambda_4 P_1(t) \qquad (9.5)$$

$$P_6'(t) = -\mu_5 P_6(t) + (1-c)\lambda_5 P_1(t) \qquad (9.6)$$

$$P_7'(t) = -\mu_2 P_7(t) + (1-c)\lambda_2 P_1(t) \qquad (9.7)$$

$$P_8'(t) = -\mu_2 P_8(t) + (1-c)\lambda_2 P_2(t) \qquad (9.8)$$

$$P_9'(t) = -\mu_1 P_9(t) + (1-c)\lambda_1 P_2(t) \qquad (9.9)$$

$$P_{10}'(t) = -\mu_4 P_{10}(t) + (1-c)\lambda_4 P_2(t) \qquad (9.10)$$

$$P_{11}'(t) = -\mu_5 P_{11}(t) + (1-c)\lambda_5 P_2(t) \qquad (9.11)$$

$$P_{12}'(t) = -\mu_1 P_{12}(t) + (1-c)\lambda_1 P_4(t) \qquad (9.12)$$

$$P_{13}'(t) = -\mu_2 P_{13}(t) + (1-c)\lambda_2 P_4(t) \qquad (9.13)$$

$$P_{14}'(t) = -\mu_3 P_{14}(t) + (1-c)\lambda_3 P_4(t) \qquad (9.14)$$

$$P_{15}'(t) = -\mu_4 P_{15}(t) + (1-c)\lambda_4 P_4(t) \qquad (9.15)$$

$$P_{16}'(t) = -\mu_5 P_{16}(t) + (1-c)\lambda_5 P_4(t) \qquad (9.16)$$

$$P_{17}'(t) = -\mu_2 P_{17}(t) + (1-c)\lambda_2 P_3(t) \qquad (9.17)$$

$$P_{18}'(t) = -\mu_3 P_{18}(t) + (1-c)\lambda_3 P_3(t) \qquad (9.18)$$

$$P_{19}'(t) = -\mu_4 P_{19}(t) + (1-c)\lambda_4 P_3(t) \qquad (9.19)$$

$$P_{20}'(t) = -\mu_5 P_{20}(t) + (1-c)\lambda_5 P_3(t) \qquad (9.20)$$

where

$$K = \lambda_1 c + (1 - c)(\lambda_2 + \lambda_4 + \lambda_5) + \lambda_3 c$$

$$L = \mu_1 + (1 - c)(\lambda_1 + \lambda_2 + \lambda_4 + \lambda_5) + \lambda_3 c$$

$$M = \mu_3 + \lambda_1 c + (1 - c)(\lambda_2 + \lambda_3 + \lambda_4 + \lambda_5)$$

$$N = \mu_1 + \mu_3 + (1 - c)(\lambda_1 + \lambda_2 + \lambda_3 + \lambda_4 + \lambda_5)$$

with initial conditions: $P_i(t) = \begin{cases} 1, & \text{if } i = 1 \\ 0, & \text{otherwise} \end{cases}$ (9.21)

The system's fuzzy availability is calculated by using the following equation:

$$A(t) = P_1(t) + 0.5P_2(t) + 0.5P_3(t) + 0.25P_4(t) \tag{9.22}$$

9.4 RESULTS AND DISCUSSION

By using Eq. (9.22), the fuzzy availability of the PET bottle hot drink filling system is calculated. Also, the effect of various failure rates, repair rates and system coverage factor on the fuzzy availability is obtained as shown in Tables 9.1–9.5.

9.4.1 EFFECT ON FUZZY AVAILABILITY OF THE SYSTEM DUE TO FAILURE AND REPAIR RATES OF THE BLOW MOLDING MACHINE

Failure rates of the blow molding machine are varied as $\lambda_1 = 0.004, 0.006, 0.008, 0.01$ and the value of all the other parameters are kept fixed as $\lambda_2 = 0.0035$, $\lambda_3 = 0.003$, $\lambda_4 = 0.0008$, $\lambda_5 = 0.005$, $\mu_1 = 0.025$, $\mu_2 = 0.015$, $\mu_3 = 0.01$, $\mu_4 = 0.02$ and $\mu_5 = 0.01$. At different values of c ranging from $c = 0$ to 1 and the above-mentioned failure and repair rates, the system's fuzzy availability is obtained and is mentioned in Table 9.1. The table shows that with the increase in the time from 20 to 60 months, the fuzzy availability of the system decreases in the range of 1.08%–12.2% with a corresponding increase in failure rate, and with increases in the value of c, the fuzzy availability increases in the range of 2.5%–14.7% for various sets of failure rate and time.

Repair rates of the blow molding machine are varied as $\mu_1 = 0.025, 0.05, 0.075, 0.1$ and the value of all the other parameters are kept fixed as $\lambda_1 = 0.004$, $\lambda_2 = 0.0035$, $\lambda_3 = 0.003$, $\lambda_4 = 0.0008$, $\lambda_5 = 0.005$, $\mu_2 = 0.015$, $\mu_3 = 0.01$, $\mu_4 = 0.02$ and $\mu_5 = 0.01$. At different values of c ranging from $c = 0$ to 1 and the above-mentioned failure and repair rates, the system's fuzzy availability is obtained and is mentioned in Table 9.1. The table shows that with the increase in the time from 20 to 60 months, the fuzzy availability of the system increases in the range of 0.01%–0.05% with a

TABLE 9.1

Effect of Failure and Repair Rates of the Blow Molding Machine on System's Fuzzy Availability

Time (in months)	System Coverage Factor (c)	Failure Rates of Blow Molding Machine (λ_1)				Repair Rates of Blow Molding Machine (μ_1)			
		0.004	0.006	0.008	0.01	0.025	0.05	0.075	0.1
0	$c=0.0$	1.0000	1.0000	1.0000	1.0000	1.0000	1.0000	1.0000	1.0000
20		0.8495	0.8495	0.8495	0.8495	0.8495	0.8495	0.8495	0.8495
40		0.7522	0.7522	0.7522	0.7522	0.7522	0.7522	0.7522	0.7522
60		0.6890	0.6890	0.6890	0.6890	0.6890	0.6890	0.6890	0.6890
0	$c=0.2$	1.0000	1.0000	1.0000	1.0000	1.0000	1.0000	1.0000	1.0000
20		0.8656	0.8621	0.8585	0.8548	0.8656	0.8657	0.8657	0.8657
40		0.7749	0.7686	0.7622	0.7558	0.7749	0.7750	0.7751	0.7751
60		0.7129	0.7045	0.6960	0.6875	0.7129	0.7132	0.7133	0.7133
0	$c=0.4$	1.0000	1.0000	1.0000	1.0000	1.0000	1.0000	1.0000	1.0000
20		0.8823	0.8752	0.8681	0.8609	0.8823	0.8824	0.8825	0.8825
40		0.7991	0.7864	0.7738	0.7612	0.7991	0.7993	0.7994	0.7994
60		0.7393	0.7222	0.7054	0.6889	0.7393	0.7397	0.7398	0.7398
0	$c=0.6$	1.0000	1.0000	1.0000	1.0000	1.0000	1.0000	1.0000	1.0000
20		0.8996	0.8889	0.8784	0.8679	0.8996	0.8997	0.8997	0.8998
40		0.8250	0.8058	0.7871	0.7689	0.8250	0.8252	0.8253	0.8253
60		0.7685	0.7425	0.7175	0.6934	0.7685	0.7688	0.7689	0.7690
0	$c=0.8$	1.0000	1.0000	1.0000	1.0000	1.0000	1.0000	1.0000	1.0000
20		0.9176	0.9033	0.8894	0.8757	0.9176	0.9176	0.9176	0.9176
40		0.8527	0.8270	0.8024	0.7788	0.8527	0.8528	0.8528	0.8528
60		0.8008	0.7657	0.7327	0.7017	0.8008	0.8009	0.8009	0.8009
0	$c=1$	1.0000	1.0000	1.0000	1.0000	1.0000	1.0000	1.0000	1.0000
20		0.9361	0.9183	0.9011	0.8845	0.9361	0.9361	0.9361	0.9360
40		0.8824	0.8502	0.8199	0.7914	0.8824	0.8823	0.8822	0.8822
60		0.8366	0.7922	0.7515	0.7140	0.8366	0.8363	0.8362	0.8362

corresponding increase in repair rate, and with increases in the value of c, the fuzzy availability increases in the range of 8.6%–14.72% for various sets of repair rate and time.

9.4.2 EFFECT ON FUZZY AVAILABILITY OF THE SYSTEM DUE TO FAILURE AND REPAIR RATES OF THE FILLING MACHINE

Failure rates of the filling machine are varied as $\lambda_2 = 0.0035$, 0.005, 0.0075, 0.01 and the value of all the other parameters are kept fixed as $\lambda_1 = 0.004$, $\lambda_3 = 0.003$, $\lambda_4 = 0.0008$, $\lambda_5 = 0.005$, $\mu_1 = 0.025$, $\mu_2 = 0.015$, $\mu_3 = 0.01$, $\mu_4 = 0.02$ and $\mu_5 = 0.01$. At different values of c ranging from $c=0$ to 1 and above-mentioned failure and repair rates, the system's fuzzy availability is obtained and is mentioned in Table 9.2. The

TABLE 9.2

Effect of Failure and Repair Rates of the Filling Machine on System's Fuzzy Availability

Time (in months)	System Coverage Factor (c)	Failure Rates of Filling Machine (λ_2)				Repair Rates of Filling Machine (μ_2)			
		0.0035	0.005	0.0075	0.01	0.015	0.030	0.045	0.06
0	$c=0.0$	1.0000	1.0000	1.0000	1.0000	1.0000	1.0000	1.0000	1.0000
20		0.8495	0.8281	0.7938	0.7611	0.8495	0.8564	0.8621	0.8669
40		0.7522	0.7210	0.6726	0.6282	0.7522	0.7709	0.7838	0.7930
60		0.6890	0.6539	0.6009	0.5538	0.6890	0.7179	0.7350	0.7457
0	$c=0.2$	1.0000	1.0000	1.0000	1.0000	1.0000	1.0000	1.0000	1.0000
20		0.8656	0.8481	0.8198	0.7925	0.8656	0.8713	0.8759	0.8798
40		0.7749	0.7488	0.7077	0.6695	0.7749	0.7902	0.8008	0.8084
60		0.7129	0.6832	0.6373	0.5957	0.7129	0.7369	0.7511	0.7600
0	$c=0.4$	1.0000	1.0000	1.0000	1.0000	1.0000	1.0000	1.0000	1.0000
20		0.8823	0.8689	0.8470	0.8256	0.8823	0.8866	0.8902	0.8931
40		0.7991	0.7786	0.7459	0.7150	0.7991	0.8109	0.8191	0.8249
60		0.7393	0.7156	0.6783	0.6437	0.7393	0.7580	0.7691	0.7760
0	$c=0.6$	1.0000	1.0000	1.0000	1.0000	1.0000	1.0000	1.0000	1.0000
20		0.8996	0.8904	0.8753	0.8605	0.8996	0.9025	0.9049	0.9069
40		0.8250	0.8107	0.7875	0.7652	0.8250	0.8331	0.8387	0.8427
60		0.7685	0.7516	0.7246	0.6989	0.7685	0.7815	0.7892	0.7940
0	$c=0.8$	1.0000	1.0000	1.0000	1.0000	1.0000	1.0000	1.0000	1.0000
20		0.9176	0.9128	0.9050	0.8973	0.9176	0.9190	0.9202	0.9212
40		0.8527	0.8452	0.8329	0.8208	0.8527	0.8569	0.8598	0.8618
60		0.8008	0.7917	0.7770	0.7626	0.8008	0.8075	0.8115	0.8140
0	$c=1$	1.0000	1.0000	1.0000	1.0000	1.0000	1.0000	1.0000	1.0000
20		0.9361	0.9361	0.9361	0.9361	0.9361	0.9361	0.9361	0.9361
40		0.8824	0.8824	0.8824	0.8824	0.8824	0.8824	0.8824	0.8824
60		0.8366	0.8366	0.8366	0.8366	0.8366	0.8366	0.8366	0.8366

table shows that with the increase in the time from 20 to 60 months, the fuzzy availability of the system decreases in the range of 2.03%–13.52% with a corresponding increase in failure rate, and with increases in the value of c, the fuzzy availability increases in the range of 8.66%–18.28% for various sets of failure rate and time.

Repair rates of the filling machine are varied as $\mu_2=0.015$, 0.03, 0.045, 0.06 and the value of all the other parameters are kept fixed as $\lambda_1 = 0.004$, $\lambda_2 = 0.0035$, $\lambda_3 = 0.003$, $\lambda_4 = 0.0008$, $\lambda_5 = 0.005$, $\mu_1=0.025$, $\mu_3=0.01$, $\mu_4=0.02$ and $\mu_5 = 0.01$. At different values of c ranging from $c=0$ to1 and the above-mentioned failure and repair rates, the system's fuzzy availability is obtained and is mentioned in Table 9.2. The table shows that with the increase in the time from 20 to 60 months, the fuzzy availability of the system increases in the range of 0.36%–5.67% with a corresponding increase in repair rate, and with increases in the value of c, the fuzzy availability increases in the range of 6.92%–14.76% for various sets of repair rate and time.

9.4.3 EFFECT ON FUZZY AVAILABILITY OF THE SYSTEM DUE TO FAILURE AND REPAIR RATES OF THE COOLING MACHINE

Failure rates of the cooling machine are varied as $\lambda_3 = 0.003, 0.005, 0.007, 0.009$ and the value of all the other parameters are kept fixed as $\lambda_1 = 0.004$, $\lambda_2 = 0.0035$, $\lambda_4 = 0.0008$, $\lambda_5 = 0.005$, $\mu_1 = 0.025$, $\mu_2 = 0.015$, $\mu_3 = 0.01$, $\mu_4 = 0.02$ and $\mu_5 = 0.01$. At different values of c ranging from $c = 0$ to 1 and the above-mentioned failure and repair rates, the system's fuzzy availability is obtained and is mentioned in Table 9.3. The table shows that with the increase in the time from 20 to 60 months, the fuzzy availability of the system decreases in the range of 1.08%–8.38% with a corresponding increase in failure rate, and with increases in the value of c, the fuzzy availability increases in the range of 4.03%–14.76% for various sets of failure rate and time.

TABLE 9.3

Effect of Failure and Repair Rates of the Cooling Machine on System's Fuzzy Availability

Time (in months)	System Coverage Factor (c)	Failure Rates of Cooling Machine (λ_3)				Repair Rates of Cooling Machine (μ_3)			
		0.003	0.005	0.007	0.009	0.01	0.04	0.07	0.1
0	$c = 0.0$	1.0000	1.0000	1.0000	1.0000	1.0000	1.0000	1.0000	1.0000
20		0.8495	0.8495	0.8495	0.8495	0.8495	0.8495	0.8495	0.8495
40		0.7522	0.7522	0.7522	0.7522	0.7522	0.7522	0.7522	0.7522
60		0.6890	0.6890	0.6890	0.6890	0.6890	0.6890	0.6890	0.6890
0	$c = 0.2$	1.0000	1.0000	1.0000	1.0000	1.0000	1.0000	1.0000	1.0000
20		0.8656	0.8624	0.8590	0.8556	0.8656	0.8668	0.8676	0.8681
40		0.7749	0.7694	0.7636	0.7577	0.7749	0.7779	0.7795	0.7804
60		0.7129	0.7058	0.6984	0.6906	0.7129	0.7178	0.7197	0.7206
0	$c = 0.4$	1.0000	1.0000	1.0000	1.0000	1.0000	1.0000	1.0000	1.0000
20		0.8823	0.8758	0.8692	0.8626	0.8823	0.8847	0.8863	0.8874
40		0.7991	0.7882	0.7771	0.7660	0.7991	0.8053	0.8085	0.8103
60		0.7393	0.7255	0.7113	0.6970	0.7393	0.7492	0.7532	0.7550
0	$c = 0.6$	1.0000	1.0000	1.0000	1.0000	1.0000	1.0000	1.0000	1.0000
20		0.8996	0.8899	0.8802	0.8706	0.8996	0.9032	0.9056	0.9073
40		0.8250	0.8089	0.7930	0.7773	0.8250	0.8346	0.8394	0.8421
60		0.7685	0.7483	0.7284	0.7089	0.7685	0.7837	0.7896	0.7925
0	$c = 0.8$	1.0000	1.0000	1.0000	1.0000	1.0000	1.0000	1.0000	1.0000
20		0.9176	0.9046	0.8920	0.8796	0.9176	0.9223	0.9256	0.9279
40		0.8527	0.8317	0.8115	0.7921	0.8527	0.8657	0.8723	0.8760
60		0.8008	0.7748	0.7503	0.7271	0.8008	0.8214	0.8295	0.8333
0	$c = 1$	1.0000	1.0000	1.0000	1.0000	1.0000	1.0000	1.0000	1.0000
20		0.9361	0.9201	0.9046	0.8898	0.9361	0.9421	0.9463	0.9492
40		0.8824	0.8567	0.8328	0.8107	0.8824	0.8991	0.9074	0.9120
60		0.8366	0.8056	0.7778	0.7528	0.8366	0.8629	0.8731	0.8779

Repair rates of the cooling machine are varied as $\mu_3 = 0.01$, 0.04, 0.07, 0.1 and the value of all the other parameters are kept fixed as $\lambda_1 = 0.004$, $\lambda_2 = 0.0035$, $\lambda_3 = 0.003$, $\lambda_4 = 0.0008$, $\lambda_5 = 0.005$, $\mu_1 = 0.025$, $\mu_2 = 0.015$, $\mu_4 = 0.02$ and $\mu_5 = 0.01$. At different values of c ranging from $c = 0$ to 1 and the above-mentioned failure and repair rates, the system's fuzzy availability is obtained and is mentioned in Table 9.3. The table shows that with the increase in the time from 20 to 60 months, the fuzzy availability of the system increases in the range of 0.01%–0.05% with a corresponding increase in repair rate, and with increases in the value of c, the fuzzy availability increases in the range of 8.6%–14.72% for various sets of repair rate and time.

9.4.4 EFFECT ON FUZZY AVAILABILITY OF THE SYSTEM DUE TO FAILURE AND REPAIR RATES OF THE CODING MACHINE

Failure rates of the coding machine are varied as $\lambda_4 = 0.0008$, 0.0012, 0.0016, 0.002 and the value of all the other parameters are kept fixed as $\lambda_1 = 0.004$, $\lambda_2 = 0.0035$, $\lambda_3 = 0.003$, $\lambda_5 = 0.005$, $\mu_1 = 0.025$, $\mu_2 = 0.015$, $\mu_3 = 0.01$, $\mu_4 = 0.02$ and $\mu_5 = 0.01$. At different values of c ranging from $c = 0$ to 1 and the above-mentioned failure and repair rates, the system's fuzzy availability is obtained and is mentioned in Table 9.4. The table shows that with the increase in the time from 20 to 60 months, the fuzzy availability of the system decreases in the range of 0.36%–2.45% with a corresponding increase in failure rate, and with increases in the value of c, the fuzzy availability increases in the range of 8.66%–17.21% for various sets of failure rate and time.

Repair rates of the coding machine are varied as $\mu_4 = 0.02$, 0.05, 0.07, 0.1 and the value of all the other parameters are kept fixed as $\lambda_1 = 0.004$, $\lambda_2 = 0.0035$, $\lambda_3 = 0.003$, $\lambda_4 = 0.0008$, $\lambda_5 = 0.005$, $\mu_1 = 0.025$, $\mu_2 = 0.015$, $\mu_3 = 0.01$ and $\mu_5 = 0.01$. At different values of c ranging from $c = 0$ to 1 and the above-mentioned failure and repair rates, the system's fuzzy availability is obtained and is mentioned in Table 9.4. The table shows that with the increase in the time from 20 to 60 months, the fuzzy availability of the system increases in the range of 0.1%–1.3% with a corresponding increase in repair rate, and with increases in the value of c, the fuzzy availability increases in the range of 8.11%–17.21% for various sets of repair rates and time.

9.4.5 EFFECT ON FUZZY AVAILABILITY OF THE SYSTEM DUE TO FAILURE AND REPAIR RATES OF THE LABELING MACHINE

Failure rates of the labeling machine are varied as $\lambda_5 = 0.005$, 0.007, 0.001, 0.012 and the value of all the other parameters are kept fixed as $\lambda_1 = 0.004$, $\lambda_2 = 0.0035$, $\lambda_3 = 0.003$, $\lambda_4 = 0.0008$, $\mu_1 = 0.025$, $\mu_2 = 0.015$, $\mu_3 = 0.01$, $\mu_4 = 0.02$ and $\mu_5 = 0.01$. At different values of c ranging from $c = 0$ to 1 and the above-mentioned failure and repair rates, the system's fuzzy availability is obtained and is mentioned in Table 9.5. The table shows that with the increase in the time from 20 to 60 months, the fuzzy availability of the system decreases in the range of 9.95%–16.57% with a corresponding increase in failure rate, and with increases in the value of c, the fuzzy availability increases in the range of 8.66%–18.61% for various sets of failure rate and time.

TABLE 9.4
Effect of Failure and Repair Rates of the Coding Machine on System's Fuzzy Availability

Time (in months)	System Coverage Factor (c)	Failure Rates of Coding Machine (λ_4)				Repair Rates of Coding Machine (μ_4)			
		0.0008	0.0012	0.0016	0.002	0.02	0.05	0.07	0.1
0	$c=0.0$	1.0000	1.0000	1.0000	1.0000	1.0000	1.0000	1.0000	1.0000
20		0.8495	0.8440	0.8386	0.8332	0.8495	0.8522	0.8535	0.8550
40		0.7522	0.7445	0.7369	0.7295	0.7522	0.7585	0.7608	0.7629
60		0.6890	0.6807	0.6725	0.6645	0.6890	0.6976	0.7001	0.7020
0	$c=0.2$	1.0000	1.0000	1.0000	1.0000	1.0000	1.0000	1.0000	1.0000
20		0.8656	0.8612	0.8567	0.8523	0.8656	0.8678	0.8689	0.8701
40		0.7749	0.7685	0.7621	0.7559	0.7749	0.7801	0.7820	0.7837
60		0.7129	0.7059	0.6990	0.6921	0.7129	0.7201	0.7222	0.7238
0	$c=0.4$	1.0000	1.0000	1.0000	1.0000	1.0000	1.0000	1.0000	1.0000
20		0.8823	0.8789	0.8755	0.8721	0.8823	0.8840	0.8848	0.8857
40		0.7991	0.7941	0.7891	0.7842	0.7991	0.8031	0.8046	0.8059
60		0.7393	0.7337	0.7282	0.7227	0.7393	0.7450	0.7466	0.7479
0	$c=0.6$	1.0000	1.0000	1.0000	1.0000	1.0000	1.0000	1.0000	1.0000
20		0.8996	0.8973	0.8950	0.8926	0.8996	0.9008	0.9013	0.9019
40		0.8250	0.8215	0.8180	0.8146	0.8250	0.8277	0.8288	0.8297
60		0.7685	0.7645	0.7606	0.7567	0.7685	0.7724	0.7736	0.7745
0	$c=0.8$	1.0000	1.0000	1.0000	1.0000	1.0000	1.0000	1.0000	1.0000
20		0.9176	0.9164	0.9152	0.9140	0.9176	0.9181	0.9184	0.9187
40		0.8527	0.8509	0.8491	0.8472	0.8527	0.8541	0.8547	0.8551
60		0.8008	0.7987	0.7966	0.7945	0.8008	0.8028	0.8034	0.8039
0	$c=1$	1.0000	1.0000	1.0000	1.0000	1.0000	1.0000	1.0000	1.0000
20		0.9361	0.9361	0.9361	0.9361	0.9361	0.9361	0.9361	0.9361
40		0.8824	0.8824	0.8824	0.8824	0.8824	0.8824	0.8824	0.8824
60		0.8366	0.8366	0.8366	0.8366	0.8366	0.8366	0.8366	0.8366

Repair rates of the labeling machine are varied as $\mu_5 = 0.01, 0.04, 0.07, 0.1$ and the value of all the other parameters are kept fixed as $\lambda_1 = 0.004, \lambda_2 = 0.0035, \lambda_3 = 0.003,$ $\lambda_4 = 0.0008, \lambda_5 = 0.005, \mu_1 = 0.025, \mu_2 = 0.015, \mu_3 = 0.01, \mu_4 = 0.02$. At different values of c ranging from $c=0$ to 1 and above-mentioned failure and repair rates, the system's fuzzy availability is obtained and is mentioned in Table 9.5. The table shows that with the increase in the time from 20 to 60 months, the fuzzy availability of the system increases in the range of 0.8%–12.58% with a corresponding increase in repair rate, and with increases in the value of c, the fuzzy availability increases in the range of 2.18%–14.76% for various sets of repair rates and time.

Constant parameters: $\lambda_2 = 0.0035, \lambda_3 = 0.003, \lambda_4 = 0.0008, \lambda_5 = 0.005, \mu_2 = 0.015,$ $\mu_3 = 0.01, \mu_4 = 0.02, \mu_5 = 0.01$

TABLE 9.5

Effect of Failure and Repair Rates of the Labeling Machine on System's Fuzzy Availability

Time (in months)	System Coverage Factor (c)	Failure Rates of Labeling Machine (λ_5)				Repair Rates of Labeling Machine (μ_5)			
		0.005	0.007	0.01	0.012	0.01	0.04	0.07	0.1
0	$c = 0.0$	1.0000	1.0000	1.0000	1.0000	1.0000	1.0000	1.0000	1.0000
20		0.8495	0.8197	0.7770	0.7500	0.8495	0.8688	0.8821	0.8915
40		0.7522	0.7066	0.6442	0.6062	0.7522	0.8038	0.8296	0.8438
60		0.6890	0.6354	0.5647	0.5233	0.6890	0.7693	0.8002	0.8148
0	$c = 0.2$	1.0000	1.0000	1.0000	1.0000	1.0000	1.0000	1.0000	1.0000
20		0.8656	0.8412	0.8059	0.7832	0.8656	0.8813	0.8921	0.8997
40		0.7749	0.7367	0.6836	0.6506	0.7749	0.8172	0.8383	0.8499
60		0.7129	0.6674	0.6059	0.5689	0.7129	0.7794	0.8047	0.8166
0	$c = 0.4$	1.0000	1.0000	1.0000	1.0000	1.0000	1.0000	1.0000	1.0000
20		0.8823	0.8635	0.8361	0.8184	0.8823	0.8943	0.9025	0.9082
40		0.7991	0.7691	0.7266	0.6998	0.7991	0.8317	0.8479	0.8567
60		0.7393	0.7030	0.6526	0.6216	0.7393	0.7909	0.8104	0.8195
0	$c = 0.6$	1.0000	1.0000	1.0000	1.0000	1.0000	1.0000	1.0000	1.0000
20		0.8996	0.8868	0.8679	0.8555	0.8996	0.9077	0.9133	0.9171
40		0.8250	0.8040	0.7738	0.7544	0.8250	0.8473	0.8584	0.8644
60		0.7685	0.7426	0.7059	0.6826	0.7685	0.8041	0.8175	0.8237
0	$c = 0.8$	1.0000	1.0000	1.0000	1.0000	1.0000	1.0000	1.0000	1.0000
20		0.9176	0.9110	0.9012	0.8947	0.9176	0.9216	0.9245	0.9264
40		0.8527	0.8417	0.8255	0.8150	0.8527	0.8642	0.8698	0.8729
60		0.8008	0.7869	0.7667	0.7536	0.8008	0.8193	0.8262	0.8294
0	$c = 1$	1.0000	1.0000	1.0000	1.0000	1.0000	1.0000	1.0000	1.0000
20		0.9361	0.9361	0.9361	0.9361	0.9361	0.9361	0.9361	0.9361
40		0.8824	0.8824	0.8824	0.8824	0.8824	0.8824	0.8824	0.8824
60		0.8366	0.8366	0.8366	0.8366	0.8366	0.8366	0.8366	0.8366

Constant parameters: $\lambda_1 = 0.004, \lambda_3 = 0.003, \lambda_4 = 0.0008, \lambda_5 = 0.005, \mu_1 = 0.025, \mu_3 = 0.01, \mu_4 = 0.02, \mu_5 = 0.001$

Constant parameters: $\lambda_1 = 0.004, \lambda_2 = 0.0035, \lambda_4 = 0.0008, \lambda_5 = 0.005, \mu_1 = 0.025, \mu_2 = 0.015, \mu_4 = 0.02, \mu_5 = 0.01$

Constant parameters: $\lambda_1 = 0.004, \lambda_2 = 0.0035, \lambda_3 = 0.003, \lambda_5 = 0.005, \mu_1 = 0.025, \mu_2 = 0.015, \mu_3 = 0.01, \mu_5 = 0.01$

Constant parameters: $\lambda_1 = 0.004, \lambda_2 = 0.0035, \lambda_3 = 0.003, \lambda_4 = 0.0008, \mu_1 = 0.025, \mu_2 = 0.015, \mu_3 = 0.01, \mu_5 = 0.02$

9.5 CONCLUSION

Analysis of fuzzy availability of PET bottle hot drink filling system before failure can be useful in improving the production and quality of the system. The detailed analysis of Tables 9.1–9.5 indicates that subsystem T, i.e. the labeling machine, is most critical as it highly affects the fuzzy availability of the system. Corresponding to various values of system coverage factor, the effect on the fuzzy availability of the system due to failure rate and repair rate is also presented in the tables. Hence, a proper care of the subsystem T should be taken by the management for better performance of the system as it is the most critical subsystem. After that, subsystems Q, P, R and S should be prioritized.

REFERENCES

1. Kai-Yuan C. Fuzzy reliability theories. *Fuzzy Sets and Systems*. 1991;40(3):510–1.
2. Kai-Yuan C, Chuan-Yuan W, Ming-Lian Z. Fuzzy reliability modeling of gracefully degradable computing systems. *Reliability Engineering & System Safety*. 1991; 33(1):141–57.
3. Zadeh, L.A. Fuzzy sets. *Information and Control*. 1965;8:338–53.
4. Zadeh L.A. Probability measures of fuzzy events. *Journal of Mathematical Analysis and Applications*. 1968;23(2):421–7.
5. Verma AK, Srividya A, Gaonkar RP. Fuzzy-reliability engineering: concepts and applications. *Narosa*. 2006.
6. Cai KY, Wen CY, Zhang ML. Fuzzy states as a basis for a theory of fuzzy reliability. *Microelectronics Reliability*. 1993;33(15):2253–63.
7. Chowdhury SG, Misra KB. Evaluation of fuzzy reliability of a non-series parallel network. *Microelectronics Reliability*. 1992;32(1–2):1–4.
8. Ke JC, Huang HI, Lin CH. Fuzzy analysis for steady-state availability: A mathematical programming approach. *Engineering Optimization*. 2006;38(8):909–21.
9. Modgil V. Mathematical modeling and time dependent availability analysis of polyethylene terephthalate bottle hot drink filling system: A case study. *Turkish Journal of Computer and Mathematics Education (TURCOMAT)*. 2021;12(10):1822–9.
10. Kumar A, Saini M. Fuzzy availability analysis of a marine power plant. *Materials Today: Proceedings*. 2018;5(11):25195–202.
11. Ouyang Z, Liu Y, Ruan SJ, Jiang T. An improved particle swarm optimization algorithm for reliability-redundancy allocation problem with mixed redundancy strategy and heterogeneous components. *Reliability Engineering & System Safety*. 2019;181:62–74.
12. Chhoker PK, Nagar A. Mathematical modeling and fuzzy availability analysis of stainless steel utensil manufacturing unit in steady state: A case study. *International Journal of System Assurance Engineering and Management*. 2015;6(3):304–18.
13. Kai-Yuan C, Chuan-Yuan W, Ming-Lian Z. Fuzzy reliability modeling of gracefully degradable computing systems. *Reliability Engineering & System Safety*. 1991; 33(1):141–57.

Index